Mineral Micro-Geochemistry Constraints on Petrogenesis and Genesis of Gold Deposit

Yayun Liang

Mineral Micro-Geochemistry Constraints on Petrogenesis and Genesis of Gold Deposit

Springer

Yayun Liang
School of Civil and Resource Engineering
University of Science and Technology Beijing
Beijing, China

ISBN 978-981-16-1021-9   ISBN 978-981-16-1022-6 (eBook)
https://doi.org/10.1007/978-981-16-1022-6

This Springer imprint is published by the registered company Springer Nature Singapore Pte Ltd.
The registered company address is: 152 Beach Road, #21-01/04 Gateway East, Singapore 189721, Singapore

Acknowledgments

This book was financially supported by the National Natural Science Foundation of China (41802077 and 41230311), the Fundamental Research Funds for the Central Universities (FRF-TP-19-023A1 and 53200759380), and China Postdoctoral Science Fund (2018M631538). We are grateful to Prof. Jun Deng, China University of Geoscience Beijing (CUGB), for his comments and assistance with funding support in the study. We extend our sincere thanks to Prof. Qingfei Wang and Xuefei Liu, CUGB, for offering advice on this research. We are grateful to Drs. Rui Zhao, Lin Yang, Yao Ma, and Yi Yang for their help during field researching and analyzing, and to editors for a lot of works and comments on this book.

Contents

Chapter 1
Introduction

The structure and elemental composition of minerals can reflect the environmental characteristics of the parent magma, retaining detailed evidence of magma evolution and crystal growth (Pilet et al. 2002; Gao et al. 2004; Zhang et al. 2007; Streck 2008; Ronick and Renno 2010; Li et al. 2013; Deng et al. 2014; Hou et al. 2015; Liu et al. 2015). In volcanic rocks, the phenocryst composition and zoning characteristics can record the magma evolution process in the subvolcanic magma chamber before the eruption (Nakagawa et al. 2002). Clinopyroxene and olivine form in the early separation and crystallisation stages of mantle-derived magma, inheriting the original characteristics of the parent magma and magma source area (Muravyeva et al. 2014). In particular, clinopyroxenes in alkaline mafic volcanic rocks have varying structures and geochemical characteristics, reflecting their origin from different parent magmas (e.g. Neumann et al. 1999; Pilet et al. 2002; Gao et al. 2008; Ronick and Renno 2010; Jankovics et al. 2016), and this type of clinopyroxene inherits important information such as open or closed magma processes, magma component evolution, and magma rise processes (Dai et al. 2015; Jankovics et al. 2016). When studying the magma mixing and source contamination, especially in case of mixed end-members that have similar isotope or trace element compositions, the accuracy of the whole rock analyses is much lower than that of in situ analyses of a single mineral. For example, the high-magnesium andesite from Mount Shasta in the United States can be interpreted to be derived from the primitive mantle according to its whole-rock data; however, its clinopyroxene phenocryst unbalanced composition and zoning structure indicate that it is derived from the mixing of felsic and mafic magma (Streck 2008). The composition of clinopyroxene formed in the early stage of magma evolution retains more primitive information than the whole rock composition. Therefore, we can more accurately explore the magma evolution process and the nature of the source by studying clinopyroxene phenocrysts in the mafic-ultramafic rock.

The analysis of H_2O content in mafic magma can provide an effective method for determining the enrichment mechanism of the mantle source (Dixon et al. 2002; Xia et al. 2013). Because different end-members contain different H_2O contents and

Y. Liang, *Mineral Micro-Geochemistry Constraints on Petrogenesis and Genesis of Gold Deposit*, https://doi.org/10.1007/978-981-16-1022-6_1

H_2O/incompatible element ratios (e.g., H_2O/Ce; Dixon et al. 2002; Liu et al. 2015), the H_2O content of the primitive magma can effectively reveal the nature of the mantle source area. Experimental studies have shown that subducted oceanic plates and marine sediments have relatively high H_2O content and high H_2O/Ce ratios (Alt and Teagle 2003). In the continental subduction environment, the subducted continental lower crust and eclogite have lower H_2O content and lower H_2O/Ce ratios (Katayama et al. 2006; Xia et al. 2006). Therefore, the H_2O content of magma (i.e., H_2O/Ce) can be used to identify whether the source area has experienced the contribution of subducted ocean or continental crustal material. The H_2O content in the original magma of basalt can be calculated by the partition coefficient of H_2O between the mineral phenocryst and the melt (O'Leary et al. 2010), because the nominally anhydrous minerals (such as clinopyroxene; Cpx) can contain significant amounts of hydroxyl (OH) in their crystal structure (Hirschmann et al. 2005; Meng et al. 2009).

North China Craton is a typical ancient craton in the earth that finished its formation during the Neoarchean to Paleoproterozoic, and has also been recognized as having undergone destruction during the Mesozoic, accompanying with large scale of magmatism and mineralization (Zhai et al. 2000; Tang and Santosh 2018). Geochemical and geophysical data have shown that the lithospheric thickness of the NCC's eastern block varied from ~200 km during the early Palaeozoic to less than 80 km during the Mesozoic–Cenozoic (Fan et al. 2000; Wu et al. 2005; Zhang et al. 2008). During the same time period, large-scale mafic to felsic magmatism and mineralisation occurred throughout the eastern NCC (Deng and Wang et al. 2016; Wang et al. 2016; Huang et al. 2017; Zhu et al. 2017). Although a lot of study focus on the Mesozoic geodynamic setting and mineralization of the eastern NCC, however, the timing and mechanism of the destruction and mineralization are still poorly understood. Jiaodong Peninsula, situated at the eastern margin of the North China Craton, presents three stages of late Mesozoic granitic intrusions, including Late Jurassic granites, Early Cretaceous granodiorites and latest Early Cretaceous granites, Cretaceous dyke swarms and large scale of gold mineralization (Goss et al. 2010; Wang et al. 2016; Yang et al. 2016; Deng et al. 2018). Many previous studies—including those addressing the geological and geochemical characteristics as well as structural characteristics, ages, materials, and fluids—have been conducted on the Jiaodong gold deposits (Deng et al. 2002; Qiu et al. 2002; Fan et al. 2003; Mao et al. 2008; Li and Santosh 2014). These studies have estimated that the main ore-forming age of the Jiaodong gold deposit is ~122 Ma (Li et al. 2006; Goldfarb and Santosh 2014; Deng et al. 2020). Previous studies have also assumed that ore-forming fluids of the Jiaodong gold province are mixed sources, but have obvious mantle-derived fluids (Mao et al. 2008; Deng et al. 2015). Moreover, the peak period of mantle-derived mafic dyke in the Jiaodong Peninsula is between 115 and 130 Ma, which is similar to the Mesozoic gold metallogenic age (~120 Ma) (Goldfarb and Santosh 2014). A large number of mafic dykes and gold deposits are close symbiosis in the gold deposits (Tan et al. 2012, 2015; Ma et al. 2014; Yang and Santosh 2015; Li et al. 2016). Therefore, the Early Cretaceous mafic dykes at Jiaodong are closely related to large-scale gold mineralization. However, some studies assume that the Jiaodong gold deposit and its symbiotic mafic dyke may only have the same source area, and that the mafic dyke did not provide ore-forming materials (Yang et al. 2014). Thus, the direct relationship between the petrogenesis of the early Cretaceous mafic dyke in Jiaodong and gold mineralization still remain controversial.

Mafic dykes, which are common in extensional tectonic settings, provide insights into the composition of the mantle and the regional geodynamic setting (Rock and Groves 1988a, b; Ernsrt et al. 1995; Sun et al. 2000; Cai et al. 2015; Deng et al. 2017). At present, the international research on mafic dykes mainly focuses on the Precambrian and Mesozoic. The study of Precambrian mafic dykes mainly reveals the reconstruction of Precambrian rifts and paleocontinents, the assembling and splitting of supercontinents, and the uplift of crystalline basement (Fahrig 1987; Hofmann 1997; Radhakrishna et al. 2013; Pisarevsky et al. 2014; Mäkitie et al. 2014). Mesozoic mafic dykes (230~100 Ma) were mainly related to the extension of the Atlantic Ocean and the breakup of the Gondwana continent (Parker et al. 1990; Pr-piper and Reynolds 2000). In China, the study on the mafic dykes mainly focuses on the Proterozoic and older mafic dykes (e.g, the North China Craton mafic dyke swarms) (Hou et al. 2006, 2008a, b; Peng et al. 2007; Peng 2010; Wang et al. 2016). The research on the Post-Proterozoic mafic dyke swarms is mainly concentrated in Tarim Basin, Qinghai-Tibet Plateau, Qinling-Dabie Orogenic Belt, the margin of North China Craton and South Yangtze Craton (Zhai and Santosh 2013; Chen et al. 2014; Deng et al. 2014). Many researchers have used mineralogy, petrology, and geochemistry to study the petrogenesis and evolution mechanism of mafic dykes, and then establish the deep dynamic evolution mechanism, providing strong evidence for the evolution of the geological background of the study area (Liu et al. al. 2009; Krüger et al. 2013; Callegaro et al. 2013; Batki et al. 2014; Stoppa et al. 2014; Guo et al. 2014; Chen et al. 2014). In addition, many researchers have found that mafic dykes and gold deposits have closely relationship of time, space, source and genesis in the typical endogenous gold deposits, such as Sublier Province, Canada, Caledoniz, Britain, Cordillera Mountains in North America, Bruson in the Italian Alps, and Australia. In China, gold deposits in the Jiaodong Peninsula, Qinling-Dabie-Sulu orogenic belt, Yunnan and other regions also found symbiosis between a large number of mafic dykes and ore bodies (Xu et al. 2012; Ma et al. 2014). Some researchers have proposed that the symbiotic gold-rich lamprophyre provides ore-forming materials during the metallogenic process of some mesothermal hydrothermal gold deposits (Rock and Groves 1988a; Wyman and Kerrich 1988; Tan et al. 2012). However, some scholars assumed that lamprophyre was rich in gold due to late alteration rather than mantle-derived and that the close symbiosis of calc-alkaline lamprophyre and mesothermal hydrothermal gold deposits only means same source rather than genesis relationship (Zhai et al. 1996; Yang et al. 2014). Therefore, the research on the mafic dykes occurred in the gold deposit is of great significance for the study of the geological evolution and gold mineralization process (Rock 1987; Rock and Groves 1988a, b; Sun et al. 2000; Jowitt et al. 2014; Tan et al. 2012, 2015; Deng et al. 2014).

Mesozoic mafic dykes are widespread on the Jiaodong Peninsula (Tan et al. 2015; Deng et al. 2017). Mafic dykes at Jiaodong are widely distributed in the Longkou-Pingdu, Penglai-Qixia, Guocheng area, Muping-Rushan area and Rongcheng-Wendeng and other major regions (Liu et al. 2005; Ma et al. 2014; Guo et al. 2013; Liang et al. 2016). In recent years, many scholars have carried out research on Jiaodong mafic dykes through using geochronology, petrology, and geochemistry aim to explore the petrogenesis of the mafic dykes, the tectonic evolution of the eastern North China Craton and relationship between mafic dykes and gold mineralization (Guo et al. 2004, 2014; Yang et al. 2004; Tan et al. 2007, 2013; Cai et al. 2013, 2015; Deng et al. 2015a; Ma et al. 2014a, b, 2016). Researchers have carried out geochronology studies of mafic dykes through different methods, mainly including whole-rock K-Ar dating (Guo et al. 2004; Liu et al. 2004; Yang et al. 2004; Liu Shen et al. 2005), biotite Ar-Ar dating (Zhang et al. 2003), and LA-ICP-MS zircon U-Pb dating (Liu et al. 2009; Cai et al. 2013, 2015; Ma et al. 2014, 2016; Liang et al. 2016), which has revealed that the formation age of mafic dykes is between 85 and 170 Ma, but mainly concentrated between 115 and 130 Ma, which is similar to the Mesozoic gold metallogenic age (~120 Ma) in the Jiaodong area (Goldfarb and Santosh 2014). Additionally, a large number of researchers have used petrology, geochemistry, and isotope geochemistry (Sr-Nd-Pb) to explore the nature of the source and tectonic background of Jiaodong mafic dykes, however, the its genesis models remain controversial (Guo et al. 2004; Yang et al. 2004; Liu et al. 2009; Tan et al. 2013; Ma et al. 2014; Cai et al. 2013, 2015). Based on previous studies on Mesozoic mafic dykes in the Jiaodong Peninsula, in this book, the in-situ major and trace elements composition, Sr isotope and H_2O content of clinopyroxene phenocryst in the dykes are analyzed in detail, which further constrains on the nature of mantle source and magmatic evolution of mafic dykes. The source metasomatism and the mixing of magma during the upwelling process of magma are discussed in detail through the micro-geochemical characteristics of minerals. The H_2O content of primitive magma reflects the essential characteristics of the Early Cretaceous lithospheric mantle in Jiaodong. It comprehensively indicates that the lithospheric mantle of the Jiaodong Peninsula has undergone paleo-Pacific subduction metasomatism and that prolong thermal-mechanical erosion has caused the thinning of the lithospheric mantle. The high content of H_2O in mafic dykes magma indicates that volatiles, such as H_2O, are provided for large-scale gold mineralization.

In this book, Chap. 2 shows the geological setting, including North China Craton and Jiadong Peninsula. Chapter 3 lists the using of mineral micro-geochemistry on the petrogenesis. Chapter 4 displays using of mineral micro-geochemistry on the genesis of gold deposit.

References

Alt, J.C., and D.A.H. Teagle. 2003. Hydrothermal alteration of upper oceanic crust formed at a fast-spreading ridge: Mineral, chemical, and isotopic evidence from ODP Site 801. *Chemical Geology* 201: 191–211.

Batki, A., P. Elemér, D. Gábor, and S. Alasdair. 2014. Petrogenetic significance of ocellar camptonite dykes in the Ditrău Alkaline Massif, Romania: Lithos 200–201, 181–196.

Cai, Y.C., H.R. Fan, M. Santosh, F.F. Hu, K.F. Yang, and Z. Hu. 2015. Subduction-related metasomatism of the lithospheric mantle beneath the southeastern north china craton: Evidence from mafic to intermediate dykes in the northern sulu orogen. *Tectonophysics* 659: 137–151.

Cai, Y.C., H.R. Fan, M. Santosh, X. Liu, F.F. Hu, K.F. Yang, T.G. Lan, Y.H. Yang, and Y.S. Liu. 2013. Evolution of the lithospheric mantle beneath the southeastern North China Craton: Constraints from mafic dykes in the Jiaobei terrain. *Gondwana Research* 24: 601–621.

Callegaro, S., Andreamarzoli, and H. Bertrand et al. 2013. Upper and lower crust recycling in the source of CAMP basaltic dykes from southeastern North America. *Earth and Planetary Science Letters* 376: 186–199.

Chen, N.H., J.J. Dong, and S.F. Yang. 2014. Restoration of geometry and emplacement mode of the Permian mafic dyke swarms in Keping and its adjacent areas of the Tarim Block, NW China. *Lithos* 204: 73–82.

Dai, L.Q., Z.F. Zhao, Y.F. Zheng, and J. Zhang. 2015. Source and magma mixing processes in continental subduction factory: Geochemical evidence from postcollisional mafic igneous rocks in the Dabie orogen. *Geochemistry Geophysics Geosystems* 16: 659–680.

Deng, J., X.F. Liu, Q.F. Wang, and R.G. Pan. 2015. Origin of the Jiaodong-type Xinli gold deposit, Jiaodong Peninsula, China: Constraints from fluid inclusion and C-D–O–S–Sr isotope compositions: Ore Geol. *Review* 65: 674–686.

Deng, J., X.F. Liu, Q.F. Wang, Y. Dilek, and Y.Y. Liang. 2017. Isotopic characterization and petrogenetic modeling of the early cretaceous mafic diking-lithospheric extension in the North China Craton, Eastern Asia. *Geological Society of America Bulletin* 129: 1379–1407.

Deng, J., Q.F. Wang, G.J. Li, and M. Santosh. 2014. Cenozoic tectonomagmatic and metallogenic processes in the Sanjiang region, southwestern China. *Earth Science Review* 138: 268–299.

Deng, J., C.M. Wang, L. Bagas, M. Santosh, E. Yao. 2018. Crustal architecture and metallogenesis in the south-eastern North China Craton. *Earth Science Reviews* 182: 251–272.

Dixon, J.E., L. Leist, C. Langmuir, and J.G. Schilling. 2002. Recycled dehydrated lithosphere observed in plume in fluenced mid-ocean-ridge basalt. *Nature* 420: 385–389.

Fahrig, W.F. 1987. The tectonic settings of continental mafic dyke swarms: Failed arm and early passive margin. *Geological Association of Canada Special Paper* 34: 331–348.

Fan, H.R., M.G. Zhai, Y.H. Xie, and J.H. Yang. 2003. Ore-forming fluids associated with granite-hosted gold mineralization at the Sanshandao deposit, Jiaodong gold province, China. *Mineralium Deposita* 38: 739–750.

Fan, W.M., H.F. Zhang, J. Baker, K.E. Jarvis, P.R.D. Mason, and M.A. Menzies. 2000. On and off the North China Craton: Where is the Archaean keel? *Journal of Petrology* 41: 933–950.

Gao, S., R.L. Rudnick, H.L. Yuan, X.M. Liu, Y.S. Liu, W.L. Xu, W.L. Ling, J. Ayers, X.C. Wang, and Q.H. Wang. 2004. Recycling lower continental crust in the North China Craton. *Nature* 432: 892–897.

Goldfarb, R.J., and M. Santosh. 2014. The dilemma of the Jiaodong gold deposits: Are they unique?: Geosci. *Front.* 2: 139–153.

Goss, S.C., S.A. Wilde, F.Y. Wu, J.H. Yang. 2010. The age, isotopic signature and significance of the youngest Mesozoic granitoids in the Jiaodong Terrane, Shandong Province, North China Craton. *Lithos* 120: 309–326.

Guo, F., W.M. Fan, C.W. Li, C.Y. Wang, H.X. Li, L. Zhao, and J.Y. Li. 2014. Hf-Nd-O isotopic evidence for melting of recycled sediments beneath the Sulu Orogen, North China. *Chemical Geology* 381: 243–258.

Guo, F., W.M. Fan, Y.J. Wang, and M. Zhang. 2004. Origin of early Cretaceous calc-alkaline lamprophyres from the Sulu orogen in eastern China: implications for enrichment processes beneath continental collisional belt. *Lithos* 78: 291–305.

Guo, P., M. Santosh, and S.R. Li. 2013. Geodynamics of gold metallogeny in the Shandong Province, NE China: an integrated geological, geophysical and geochemical perspective. *Gondwana Research* 24: 1172–1202.

Hirschmann, M.M., C. Aubaud, and A.C. Withers. 2005. Storage capacity of H_2O in nominally anhydrous minerals in the upper mantle. *Earth and Planetary Science Letters* 236: 167–181.

Hofmann, A.W. 1997. Mantle geochemistry: The message from oceanic volcanism. *Nature* 385: 219–229.

Hou, G.T., J.H. Li, M.H. Yang, W. Yao, C. Wang, and Y. Wang. 2008a. Geochemical constraints on the tectonic environment of the late paleoproterozoic mafic dyke swarms in the north china craton. *Gondwana Research* 13 (1): 103–116.

Hou, G.T., Y. Liu, and J. Li. 2006. Evidence for ~1.8 Ga extension of the Eastern Block of the North China Craton from SHRIMP U-Pb dating of mafic dyke swarms in Shandong Province. *Journal of Asian Earth Sciences* 27(4): 392–401.

Hou, G.T., M. Santosh, and X.L. Qian. 2008b. Configuration of the Late Paleoproterozoic supercontinent Columbia: Insights from radiating mafic dyke swarms. *Gondwana Research* 14: 395–409.

Hou, T., Z.C. Zhang, J.K. Keiding, and I.V. Veksler. 2015. Petrogenesis of the ultrapotassic Fanshan intrusion in the North China Craton: implications for lithospheric mantle metasomatism and the origin of apatite ores. *Journal of Petrology* 5: 893–918.

Huang, F., J.F. Xu, Y.S. Liu, J. Li, J.L. Chen, and X.Y. Li. 2017. Re-Os isotope evidence from Mesozoic and Cenozoic basalts for secular evolution of the mantle beneath the North China Craton. *Contributions to Mineralogy and Petrology* 172: 28.

Jankovics, M.É., Z. Taracsák, G. Dobosi, A. Embey-Isztin, A. Batki, S. Harangi, and C.A. Hauzenberger. 2016. Clinopyroxene with diverse origins in alkaline basalts from the western Pannonian Basin: Implications from trace element characteristics. *Lithos* 262: 120–134.

Jowitt, S.M., M.C. Williamson, R.E. Ernst. 2014. Geochemistry of the 130 to 80 Ma Canadian High Arctic large igneous province (HALIP) event and implications for Ni-Cu-PGE prospectivity. *Economic Geology* 109: 281–307.

Katayama, I., S. Nakashima, and H. Yurimoto. 2006. Water content in natural eclogite and implication for water transport into the deep upper mantle. *Lithos* 86: 245–259.

Krüger, J.C., L.H. Romerr, and H. Kämpf. 2013. Late Cretaceous ultramafic lamprophyres and carbonatites from the Delitzsch Complex, Germany: Chem. *Geology* 353: 140–150.

Li, L., S.R. Li, M. Santosh, Q. Li, Y. Gu, W.J. Lü. 2016. Dyke swarms and their role in the genesis of world-class gold deposits: Insights from the Jiaodong peninsula, China. *Journal of Asian Earth Sciences*.

Li, J.W., P. Vasconcelos, M.F. Zhou, X.F. Zhao, and C.Q. Ma. 2006. Geochronology of the Pengjiakuang and Rushan gold deposits, eastern Jiaodong Gold Province, northeastern China: Implications for regional mineralization and geodynamic setting. *Economic Geology* 101: 1023–1038.

Li, X.W., X.X. Mo, X.H. Yu, Y. Ding, X.F. Huang, and P. Wei. 2013. Petrology and geochemistry of the early Mesozoic pyroxene andesites in the Maixiu area, west Qinling, china: products of subduction or syn-collision? *Lithos* 172–173: 158–174.

Liang, Y.Y., X.F. Liu, C. Qin, Y. Li, J. Chen, J.Y. Jiang. 2016. Petrogenesis of Early Cretaceous mafic dikes in southeastern Jiaolai basin Jiaodong Peninsula China. *International Geology Review* 2: 131–150.

Liu, S., R.Z. Hu, J.H. Zhao. 2004. K-Ar Geochronology of Mesozoic Mafic Dikes in Shandong Province, Eastern China: Implications for Crustal Extension. *Acta Geologica Sinica* 78: 1207–1213.

Liu, S., R.Z. Hu, J.H. Zhao, C.X. Feng, H. Zhong, J.J. Cao, and D.N. Shi. 2005. Geochemical characteristics and petrogenetic investigation of the late Mesozoic lamprophyres of Jiaobei, Shandong Province. *Acta Petrologica Sinica* 21: 947–958 (in Chinese with English abstract).
Liu, J., Q.K. Xia, E. Deloule, H. Chen, and M. Feng. 2015. Recycled oceanic crust and marine sediment in the source of alkali basalts in Shandong, eastern China: Evidence from magma water content and oxygen isotopes. *Journal of Geophysical Research: Solid Earth* 120: 8281–8303.
Liu, S., R. Hu, S. Gao, C. Feng, B. Yu, G. Feng, Y. Qi, T. Wang, and I.M. Coulson. 2009. Petrogenesis of Late Mesozoic mafic dykes in the Jiaodong Peninsula, eastern North China Craton and implications for the foundering of lower crust. *Lithos* 113: 621–639.
Ma, L., S.Y. Jiang, A.W. Hofman, Y.G. Xu, B.Z. Dai, and M.L. Hou. 2016. Rapid lithospheric thinning of the North China Craton: New evidence from cretaceous mafic dykes in the Jiaodong Peninsula. *Chemical Geology* 432: 1–15.
Ma, L., S.Y. Jiang, and A. Hofman. 2014. Lithospheric and asthenospheric sources of lamprophyres in the Jiaodong Peninsula: a consequence of rapid lithospheric thinning beneath the North China Craton? *Geochimica et Cosmochimica Acta* 124: 250–271.
Mäkitie, H., G. Data, and E. Isabirye. 2014. Petrology, geochronology and emplacement model of the giant 1.37 Ga arcuate lake victoria dyke swarm on the margin of a large igneous province in eastern Africa. *Journal of African Earth Sciences* 97: 273–296.
Mao, J.W., Y.T. Wang, H.M. Li, F. Pirajno, C.Q. Zhang, and R.T. Wang. 2008. The relationship of mantle-derived fluids to gold metallogenesisin the Jiaodong Peninsula Evidence from D-O-C-S isotope systematics. *Ore Geology Reviews* 33: 361–381.
Meng, D.W., X.L. Wu, X.Y. Fan, X. Meng, J.P. Zheng, and R. Mason. 2009. Submicron-sized fluid inclusions and distribution of hydrous components in jadeite, quartz and symplectite-forming minerals from UHP jadeite–quartzite in the Dabie Mountains, China: TEM and FTIR investigation. *Applied Geochemistry* 24: 517–526.
Muravyeva, N.S., B.V. Belyatsky, V.G. Senin, and A.V. Ivanov. 2014. Sr-Nd-Pb isotope systematics and clinopyroxene-host disequilibrium in ultra-potassic magmas from Toro-Ankole and Virunga, East-African Rift: Implications for magma mixing and source heterogeneity. *Lithos* 210: 260–277.
Nakagawa, M., K. Wada, and C.P. Wood. 2002. Mixed magmas, mush chambers and eruption triggers: Evidence from zoned clinopyroxene phenocrysts in andesitic scoria from the 1995 eruptions of Ruapehu volcano, New Zealand. *Journal of Petrology* 43: 2279–2303.
Neumann, E.R., E. Wuff-Pedersen, S.L. Simonsen, N.J. Pearson, J. Marti, and J. Mitjacila. 1999. Evidence for fractional crystallization of periodically refilled magma chambers in Tenerife, Canary Islands. *Journal of Petrology* 40: 1089–1123.
Parker, A.J., P.C. Rickwood, and D.H. Tucker. 1990. Mafic dykes and emplacement mechanisms. Routledge, 541.
Peng, P. 2010. Reconstruction and interpretation of giant mafic dyke swarms: A case study of 1.78 Ga magmatism in the North China craton. *Geological Society, London, Special Publications* 338: 163–178.
Peng, P., M.G. Zhai, J.H. Guo, T. Kusky, and T.P. Zhao. 2007. Nature of mantle source contributions and crystal differentiation in the petrogenesis of the 1.78ga mafic dykes in the central north china craton. *Gondwana Research* 12 (1-2): 29-46.
Pilet, S., J. Hernandez, and B. Villemant. 2002. Evidence for high silicic melt circulation and metasomatic events in the mantle beneath alkaline provinces: the Na-Fe-augitic green-core pyroxenes in the tertiary alkali basalts of the Cantal massif (French Massif Central). *Mineralogy and Petrology* 76: 39–62.
Pisarevsky, S.A., M.T.D. Wingate, and Z.X. Li. 2014. Age and paleomagnetism of the 1210 Ma gnowangerup–fraser dyke swarm, western Australia, and implications for late mesoproterozoic paleogeography. *Precambrian Research* 246: 1–15.
Radhakrishna, T., N.R. Krishnendu, and G. Balasubramonian. 2013. Palaeoproterozoic indian shield in the global continental assembly: evidence from the palaeomagnetism of mafic dyke swarms. *Earth-Science Reviews* 126: 370–389.

Rock, N.M.S., and D.I. Groves. 1988a. Can lamprophyres resolve the genetic controversy over mesothermal gold deposits? *Geology* 16: 538–541.

Rock, N.M.S., and D.I. Groves. 1988b. Do lamprophyres carry gold as well as diamonds? *Nature* 332: 253–255.

Ronick, R., and A.D. Renno. 2010. Clinopyroxene from basaltic rocks of the Erzgebirge-Krusne hory Mts.-implications for modelling the magmatic plumbing system. *Journal of Geosciences* 55: 231–249.

Streck, M.J. 2008. Mineral textures and zoning as evidence for open system processes. *Reviews in Mineralogy and Geochemistry* 69: 595–622.

Sun, J.G., S.X. Hu, and F. Hong. 2000. Study on the geochemistry and subduction–crust mantle interaction of the high potassium–potassium dike rocks in gold deposits concentration zone of east Shandong, China: Acta Petrol. *Sin.* 16: 401–412 (in Chinese with English abstract).

Tan, J., J. Wei, A. Audétat, and T. Pettke. 2012. Source of metals in the Guocheng gold deposit, Jiaodong Peninsula, North China Craton: Link to early Cretaceous mafic magmatism originating from Paleoproterozoic metasomatized lithospheric mantle. *Ore Geology Reviews* 48: 70–87.

Tan, J., J.H. Wei, Y.G. Li, W.J. Tan, G. Dazhao, and C. Yang. 2007. Geochemical characteristics of Late Mesozoic Dikes, Jiaodong Peninsula, North China Craton: Petrogenesis and Geodynamic Setting: Int. *Geol. Rev.* 49: 931–946.

Tan, J., J.H. Wei, W.J. Shi, B. Feng, Y.J. Li, and L.B. Fu. 2013. Origin of dyke swarms by mixing of metasomatized subcontinental lithospheric mantle-derived and lower crustal magmas in the Guocheng fault belt, Jiaodong Peninsula, North China Craton. *Geological Journal* 48: 516–530.

Tan, J., J.H. Wei, Y.J. Li, L.B. Fu, H.M. Li, W.J. Shi, and N. Tian. 2015. Origin and geodynamic significance of fault-hosted massive sulfide gold deposits from the Guocheng-Liaoshang metallogenic belt, eastern Jiaodong Peninsula: Rb–Sr dating, and H-O–S–Pb isotopic constraints: Ore Geol. *Review* 65: 687–700.

Tang, L., and M. Santosh. 2018. Neoarchean-Paleoproterozoic terrane assembly and Wilson cycle in the North China Craton: an overview from the central segment of the Trans-North China Orogen. *Earth-Science Reviews* 182: 1–27.

Wang, C.M., Y. Lu, X. He, M. Jessell, Q. Wang, and J. Zhang. 2016. The Paleoproterozoic diorite dykes in the southern margin of the North China Craton: Insight into rift-related magmatism. *Precambrian Research* 277: 26–46.

Wu, F.Y., J.Q. Lin, S.A. Wilde, X.O. Zhang, and J.H. Yang. 2005. Nature and significance of the Early Cretaceous giant igneous event in eastern China. *Earth and Planetary Science Letters* 233: 103–119.

Wyman, D., and R. Kerrich. 1988. Lamprophyres a source of gold. *Nature* 332: 209–210.

Xia, Q.K., Y. Bi, P. Li, W. Tian, X. Wei, and H.L. Chen. 2016. High water content in primitive continental flood basalts. *Scientific Reports* 6: 25416.

Xia, Q.K., J. Liu, S.C. Liu, I. Kovács, M. Feng, and L. Dang. 2013. High water content in mesozoic primitive basalts of the North China Craton and implications on the destruction of cratonic mantle lithosphere. *Earth Planetary Science Letters* 361: 85–97.

Xia, Q.K., X.Z. Yang, E. Deloule, Y.M. Sheng, and Y.T. Hao. 2006. Water in the lower crustal granulite xenoliths from Nushan, eastern China. *Journal of Geophysical Research: Solid Earth* 111: B11202.

Xu, Y.G., H.H. Zhang, H.N. Qiu, W.C. Ge, and F.Y. Wu. 2012. Oceanic crust components in continental basalts from Shuangliao, Northeast China: Derived from the mantle transition zone? *Chemical Geology*: 168–184.

Yang, D.B., W.L. Xu, G.C. Zhao, T.F. Huo, J.P. Shi, and H.T. Yang. 2016. Tectonic implications of Early Cretaceous low-Mg adakitic rocks generated by partial melting of thickened lower continental crust at the southern margin of the central North China Craton. *Gondwana Research* 38: 220–237.

Yang, J.H., S.L. Chung, M.G. Zhai, and X.H. Zhou. 2004. Geochemical and Sr-Nd-Pb isotopic compositions of mafic dykes from the Jiaodong Peninsula, China: evidence for vein-plus-peridotite melting in the lithospheric mantle. *Lithos* 73: 145–160.

Yang, L.Q., J. Deng, Z.L. Wang, and L. Zhang. 2014. Mesozoic gold metallogenic system of the Jiaodong gold province, eastern China. *Acta Geologica Sinica* 30 (9): 2447–2467 (in Chinese with English abstract).

Yang, Q.Y., and M. Santosh. 2015. Early Cretaceous magma flare-up and its implications on gold mineralization in the Jiaodong Peninsula, China. *Ore Geology Reviews* 65: 626–642.

Zhai, M.G., A. Bian, and T. Zhao. 2000. The amalgamation of the supercontinent of North China Craton at the end of Neo-Archaean and its breakup during late Palaeoproterozoic and Meso-Proterozoic. *Science in China Series D: Earth Sciences* 43 (1): 219–232.

Zhai, M.G., and M. Santosh. 2013. Metallogeny of the North China Craton: Link with secular changes in the evolving Earth. *Gondwana Research* 24: 275–297.

Zhang, H.F., S.L. Goldstein, X.H. Zhou, M. Sun, J.P. Zheng, and Y. Cai. 2008. Evolution of subcontinental lithospheric mantle beneath eastern China: Re-Os isotopic evidence from mantle xenoliths in Paleozoic kimberlites and Mesozoic basalts. *Contributions to Mineralogy and Petrology* 155: 271–293.

Zhang, H.F., J.F. Ying, G. Shimoda, N.T. Kita, Y. Morishita, J.A. Shao, and Y.J. Tang. 2007. Importance of melt circulation and crust-mantle interaction in the lithospheric evolution beneath the North China Craton: Evidence from Mesozoic basalt-borne clinopyroxene xenocrysts and pyroxenite xenoliths. *Lithos* 96: 67–89.

Zhang, L.C., S.Y. Chao, Q.D. Zeng, and G.M. Li. 2003. $^{40}Ar/^{39}Ar$ and Rb-Sr isochron dating of the gold deposits on northern margin of the Jiaolai Basin, Shandong, China. *SCIENCE IN CHINA (Series D)* 46: 708–718.

Zhai, J.P., K. Hu, J.J. Lu. 1996. A discussion on the new genetic model for lamprophyre and gold mineralization. *Mineral Deposits* 15: 80–86. (in Chinese with English abstract).

Zhu, R.X., H.F. Zhang, G. Zhu, Q.R. Meng, H.R. Fan, J.H. Yang, F.Y. Wu, Z.Y. Zhang, and T.Y. Zheng. 2017. Craton destruction and related resources. *International Journal of Earth Sciences* 106: 2233–2257.

Chapter 2
Geological Setting

Abstract North China Craton, one of the oldest cratons in China, consists of west block, Trans-North China orogen, and east block. The NCC is bordered by the Central Asian orogenic belt to the north, the Qinling–Dabie–Sulu orogenic belt to the south, and the Pacific Ocean to the east. NCC wasn't tectonically stable until Paleozoic since formed during Neoarchean to Paleoproterozoic (18 Ga ago). From the Triassic to the Jurassic period, the NCC and the Yangtze craton (YC) collided, accompanied by continental subduction and crustal thickening, forming the Qinling–Dabie–Sulu orogenic belt. During Middle Triassic (~220 Ma), the northward subduction of crust of Yangtze Craton under NCC to mantle depths of >200 km. Since the Early Jurassic, the eastern part of the NCC, including Jiaodong Peninsula, have been influenced by the westward subduction of the Pacific plate. The NCC has undergone extensive lithosphere thinning from a thickness of about 200 km to 60–80 km with the peak occurring in the Cretaceous, while the eastern region underwent lithospheric thinning and delamination, accompanied by large-scale magmatism and gold mineralization.

Keywords Eastern North China Craton · Jiaodong Peninsula · Lithospheric thinning · Gold deposit · Magmatism

2.1 North China Craton

North China Craton is a typical ancient craton with a 3.8 billion year-history marked by complex and multi-stage tectonic evolution, and covers approximately 300,000 km^2, consisting of west block, Trans-North China orogen, and east block (Zhao and Zhai 2013). During the Archean period (>3.0 Ga), the North China Craton experienced the formation of continental nuclei and microcontinents. Accretion of continental crust occurred from 2.7–2.9 Ga. During the ~2.5 Ga period, magmatism, metamorphism, and cratonisation occurred (Zhai 2010). A series of sedimentary rocks and bimodal volcanic rocks revealed that the North China Craton was uplifted during the Paleoproterozoic period (approximately 2.3–2.0 Ga) (Zhai and Santosh 2013). At ~1.85 Ga, the eastern and western blocks of the North China Craton collided and merged into the central orogenic belt, completing cratonisation (Zhao et al. 2001, 2005). From the Paleoproterozoic to Neoproterozoic, the North China

Y. Liang, *Mineral Micro-Geochemistry Constraints on Petrogenesis and Genesis of Gold Deposit*, https://doi.org/10.1007/978-981-16-1022-6_2

Craton experienced many uplifting movements and was in a relatively stable platform state from the Neoproterozoic to Paleozoic. NCC wasn't tectonically stable until Paleozoic since formed during Neoarchean to Paleoproterozoic (1.8 Ga ago) (Zhai and Santosh 2013). From the Paleozoic to Early Mesozoic, the North China Plate underwent multiple periods of orogenic activity (Chen et al. 2009; Dong et al. 2011). During the Triassic to Jurassic periods, the North China Craton and the Yangtze Craton collided and merged, accompanied by continental subduction and thickening of the crust to form the Qinling–Dabie–Sulu orogenic belt (Li et al. 1993; Dong et al. 2011), which slips to the left for approximately 500–700 km along the Tan–Lu fault zone (Zhu et al. 2010). From the Late Permian to Early Jurassic, the northern margin of the North China Craton experienced continuous compression and crustal accretion as the Paleo-Asian Ocean closed (Xiao et al. 2003). From the Late Jurassic to the Early Cretaceous, the western block of the North China Craton (primarily the Ordos plate) was relatively stable, whereas the eastern block underwent extensional tectonic activities related to lithospheric thinning and decratonisation (Zhu et al. 2010; Deng et al. 2015) (Fig. 2.2). During the Mesozoic, the eastern block of the North China Craton and even the entire eastern China experienced the westward subduction of the Pacific Plate (Li et al. 2007; Zhou and Li 2000; Jiang et al. 2010). During the Early Cretaceous, large-scale metamorphic core complexes, pull-apart basins, A-type granites, and dyke swarms were formed in the eastern block of the North China Craton, indicating that the eastern block was in a regional extensional environment during this period. Simultaneous lithospheric thinning may be related to the rollback of the subducting Pacific Plate along the eastern edge of Asia (Yang et al. 2007).

2.2 Jiaodong Gold Province

The Jiaodong Peninsula is situated at the eastern margin of the North China Craton (Fig. 2.1) and has been divided into the Jiaobei terrane and the Sulu orogenic belt by the Wulian–Yantai Fault (Fig. 2.2). The Sulu orogenic belt is primarily composed of Neoproterozoic granitic gneisses with minor coesite-bearing eclogites, quartzite, and schist (Zhang et al. 2012). The Jiaobei terrane consists of the Jiaobei uplift and the Jiaolai Basin. The Jiaobei terrane—adjacent to the Bohai Sea to the north and the Yellow Sea to the east—is a large-scale gold-mining area and one of the principal industrial gold-producing areas in China. The Precambrian high-grade metamorphic rocks in the Jiaobei uplift include the Neoarchean Jiaodong Group, the Paleoproterozoic Fenzishan and Jinshan Groups, and the Meso-Neoproterozoic Penglai Group (Zhai et al. 2000; Tang et al. 2008). The Jiaolai basin belongs to the Cretaceous pull-apart basin, composed of the early-Cretaceous Laiyang Group and the Qingshan Group, late-Cretaceous Wangshi Group, and the Cenozoic Huangxian Group and the Yaoshan Group (Tang et al. 2008). The Mesozoic granitic intrusive rocks are primarily distributed in the Jiaobei terrane and can be grouped into Late Jurassic plutons (166–146 Ma), Early Cretaceous intrusions (136–123 Ma), and latest Early

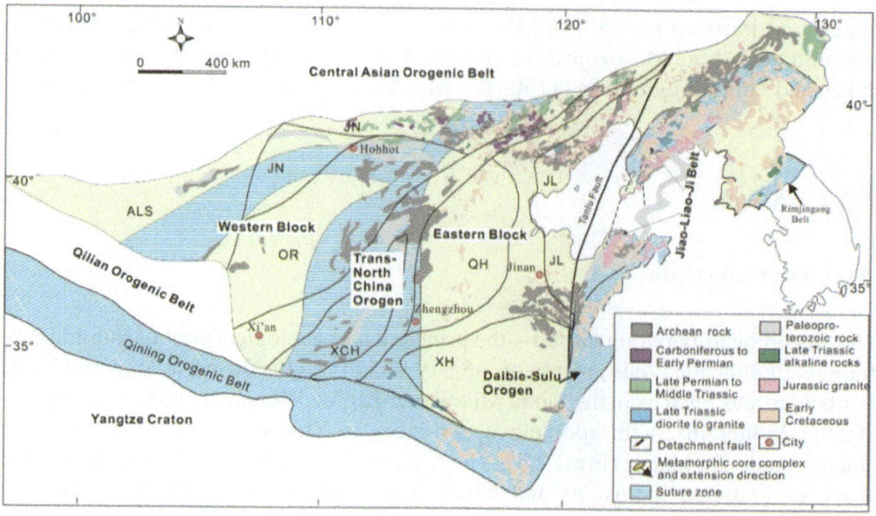

Fig. 2.1 The simple graph of North China Craton (Modified from Zhao and Zhai 2013)

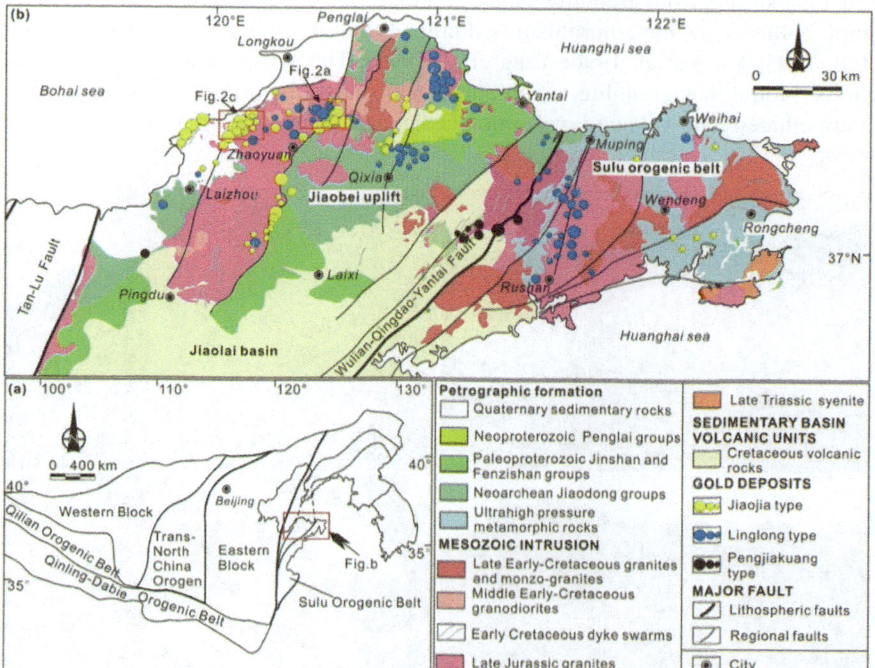

Fig. 2.2 **a** Diagram of the North China Craton with the location of Jiaodong gold province; **b** simplified geological map of the Jiaodong gold province (Modified from Yang et al. 2016)

Cretaceous intrusions (123–108 Ma) (Goss et al. 2010; Yang et al. 2012). Additionally, numerous Late Mesozoic dyke swarms were consisted of diorite dykes, dolerite dykes and lamprophyre dykes (130–110 Ma; Ma et al. 2014; Deng et al. 2017).

2.2.1 *Strata*

(1) **Archean strata**

The Neoarchean Jiaodong Group—the principal archaeological strata exposed in the Jiaodong Peninsula—has an area of ~5,519.96 km^2 and is primarily exposed in the Jiaobei terrane, such as in the Qixia Miaojiabei-Linjia, Wujiabei, and Guogezhuang-Dafang Mountains, with sporadic occurrences in Zhaoyuan Sujiadian, Longkou Quanzi, and Letukuang (Figs. 2.3). These features are primarily distributed in the shape of scattered inclusions and bands in the northeast-east (NEE)–northwest-west (NWW) direction and widely distributed in the Jingshan Group with a hidden unconformity or fault contact relationships (Chen et al. 1993). The Jiaodong Group generally suffers from granulite facies–amphibolite facies metamorphism, of which amphibolite facies metamorphism is dominant and reaches granulite facies (Chen et al. 1993; Yang et al. 1996; Yang et al. 1998). The principal lithologies include gneiss, biotite metagranulite, amphibolite, and schist; the original rocks are mafic-intermediate-felsic volcanic, volcanic clastic, and intercalated marine argillaceous

Fig. 2.3 The representative field photographs of dyke swarms in the Jiaodong Peninsula

carbonates. The SHRIMP zircon U–Pb geochronology shows that the amphibolite protolith was formed at 2530 ± 17 Ma and that metamorphism occurred at 1852 ± 37 Ma (Zhang et al. 2003). The Jiaodong Group can be divided into three groups from top to bottom, namely, the Tangjiazhuang, Qishan, and Linjiazhai formations (Yang et al. 1996). The Tangjiazhuang Formation includes biotite amphibolite schistite, amphibolite metagranulite, and biotite amphibolite gneiss; the Qishan Formation is primarily composed of amphibolite, biotite schist, and biotite metagranulite; and the Linjiazhai Formation is primarily composed of metagranulite, biotite metagranulite, and biotite amphibolite schist, which contains plagioclase amphibolite.

(2) **Proterozoic strata**

The Proterozoic strata exposed in the Jiaodong Peninsula are primarily from the Paleoproterozoic Jingshan, Fenzishan, and Neoproterozoic Penglai Groups (Fig. 2.3). The Paleoproterozoic Jingshan Group is primarily distributed in the Jiaobei terrane, such as in Laiyang Jingshan, Jingqishan, Mupingguangshan, Xiangshan, Pingdu Mingcun, Xianshan, Laixi Nanshu, Haiyang Jingshan, and Qixia Pagoda. Sporadic distributions are found in extreme locations in the Sulu orogenic belt, such as Rongcheng and Weihai. The degree of metamorphism of the Jingshan Group reaches granulite facies–amphibolite facies. The Jingshan Group underwent strong fold deformation, which often constitutes superimposed fold structures, is in concealed unconformity or fault contact with the underlying Jiaodong Group (Chen et al. 1993), and has an angular unconformity contact relationship with the overlying Fanzishan Group (Yang et al. 1996). The isotopic geochronology was mostly concentrated between 1800 and 2484 Ma. The metamorphic facies are primarily low-degree amphibolite facies, high-degree green schist facies with minor amphibolite facies. Principal lithologies include marble, schist, and metagranulite. The Jingshan Group is divided into the Lugezhuang, Yetou, and Douya Formations. The upper part of the Lugezhuang Formation is composed of silicobilotite schist, amphibolite, amphibolite, serpentinised marble, and metagranulite, whereas the lower part is composed of serpentinite, serpentinised peridotite, and serpentinised marble. The Yetou Formation is composed of diopside marble, marble, and biotite granulite. The Douya Formation consists of two dolomite quartz schists, graphite mica schists, sericite quartz schists, and biotite quartz schists (Li et al. 1993).

The Paleoproterozoic Fengzishan Group—primarily distributed in Laizhou Fenzishan, Qixia Miaohou, Penglai Jinguo Mountain, Yantai Guxian, Fushandongting, and Zhanggezhuang—has a different location and isotopic strata to and a pseudo-integrated contact relationship with the Jiaodong Group (Li et al. 1993; Yang et al. 1996). An overlapping phase transition relationship joins the upper strata of the Jingshan and Jingshan groups (Chen et al. 1993). The isotopic ages of the Fengzishan Group were mostly between 1674 and 2075 Ma. The metamorphic facies are primarily high green schist facies with low amphibolite facies and consist of marble, graphite tremolite, biotite, plagioclase amphibolite, tremolite, and shallow granules.

The Fenzishan Group is divided into two subgroups based on the characteristics of the protolith: the lower (South Fenzishan Group) and upper (North Fenzishan Group) subgroups. The lower subgroup is primarily composed of metagranulite, marble, schist, and amphibolite. The protolith is a set of ultrabasic-mafic-intermediate felsic volcanic rocks and argillaceous carbonates. The upper subgroup was primarily schist and marble. The protolith is a set of clay deposits, clastic deposits, and carbonate rocks (Li et al. 1993; Yang et al. 1996).

The Neoproterozoic Penglai Group is primarily distributed in eastern Penglai, western Fushan and Zangjiazhuang, and Nanzhuang in Qixia (Chen et al. 1993; Yan et al. 1996). According to ancient plant assemblages and stromatolite assemblages found in this group, the Penglai Group was determined to be between 600 and 700 Ma. The Penglai Group underwent low-green schist facies metamorphism and is a set of shallow metamorphic sedimentary rocks with an angular unconformity covering the Archean Jiaodong and Proterozoic Fenzishan Groups that primarily consists of phyllite, quartzite, crystalline limestone, slate, and marble. The protolith consists of a set of argillaceous carbonates, continental clastics and carbonates.

(3) **Phanerozoic strata**

The Phanerozoic strata in Jiaodong are primarily Mesozoic and Cenozoic. Mesozoic strata include the Laiyang, Qingshan, and Wangshi Groups; the Cenozoic strata include Tertiary and Quaternary sedimentary rocks.

The Mesozoic Laiyang Group is primarily distributed in the periphery and inside the Jiaolai Basin, Qixia Zangjiazhuang, western Penglai, and Muping Kongxintou and belongs to a set of sedimentary formations of river-lacustrine clastic rocks intercalated with volcanic rocks. The zircon U–Pb ages of the volcanic rocks in the Laiyang Group and the $^{40}Ar/^{39}Ar$ age of amphibole are restricted to ~130 Ma (Zhang et al. 2008). The Qingshan Group is primarily distributed in the periphery of the Jiaolai Basin, Qixia Zangjiazhuang, and Rongcheng Lidao areas and is composed of acidic-medium-basic-acidic-basic volcanic rocks and pyroclastic rocks. The eruption age of the basic volcanic rocks of the Qingshan Group is 122–113 Ma, and the eruption age of the acidic volcanic rocks is 110–98 Ma (Li et al. 2007). The Wangshi Group is primarily distributed in Jushangou, Linjiazhuang, Qixia Zhaili, northern Guanshui, and Xiangli of Fushan and primarily includes clastic rocks and a few volcanic rocks. The $^{40}Ar/^{39}Ar$ age of the hornblende in the Daxinzhuang basalt is 73.5 ± 0.3 Ma (Yan et al. 2003).

The Cenozoic Tertiary and Quaternary strata are distributed in eastern Laizhou-Huangxian, southern Pingdu, Zetou Rushan, and Chengxiang Rongcheng. The principal lithology is sandstone and sand shale, reflecting a continental sedimentary environment (Li et al. 1993; Yang et al. 1996).

2.2.2 Structure

(1) **Folds**

The folds of the Jiaodong Peninsula occur in metamorphic crystalline basements, including the Huangxian-Penglai, Sanshandao-Linglong-Muping, Zhaoyuan-Qixia-Wangtuan, Laiyang-Zetou, and Laixi-Rushan belts, which are distributed in a nearly EW direction. The basement folds developed in this area are believed to have been formed by the compression—with a N-S orientation—of the Fuping, Luliang, and Wutai movements before cratonisation (Yang et al. 1998).

(2) **Faults**

The fault zone of the Jiaodong Peninsula is oriented in an approximately northeast (NE)– north-northeast (NNE) direction and includes the Sanshandao–Cangshang, Jiaojia, Zhaoye–Pingdu, Wulian–Yantai, Muping–Jimo, Mishan, and Qingdao–Shidao secondary faults (Zhai et al. 2004; Mao et al. 2005) of the regional Tancheng–Lujiang fault (Goldfarb et al. 2001). The Tan–Lu fault is located on the western edge of the Jiaodong Peninsula and is primarily composed of four fault zones, including the Yishui–Tangtou, Chuyun–Gegou, Anqiu–Juxian, and Changyi–Dadian faults. Of these, the Anqiu–Juxian fault divides the two major blocks of Jiaodong and Luxi (Li et al. 2010). The fault zone strikes in the NEE direction (10°–20°) and has a length of >300 km and width of 20–60 km in the Shandong area.

The Sanshandao–Cangshang fault zone, located to the east of Laizhou Bay—which extends from NE and southwest ends to the Bohai Sea—has a length and width of approximately 11 km and 50–200 m, respectively, trends NE at 40°, is inclined to the SE, and has a dip angle of 35°–45°. The fault zone spreads in irregular ribbons along the strike and shows soothing wavy curves. The upper part of the fault is superimposed on the contact zone of the Jiaodong Group strata and the Linglong and Guojialing granites; the northeast section of the lower part extends into the Guojialing granite.

The Jiaojia fault zone—located approximately 15–20 km east of the Sanshandao fault—starts in Longkou Huangshanguan in the north, passes through Xincheng and Jiaojia to the south, and reaches Pinglidian farther south. The fault has a length of 50 km and a width of 30–300 m and spreads in a gentle wave shape with a trend of 20°–40°, northwest (NW) tendency, and dip angle of 30°–45°. The principal fracture surface of the Jiaojia fault is continuous with thick layers of grey–white and grey–black fault gouges. The structural alteration zone within the fault is obvious and exhibits multiple periods of activity, showing compression and torsion pro-mineralisation, tension and torsion during mineralisation, and compression and torsion post-mineralisation.

The Zhaoyuan–Pingdu fault zone starts from Pingdu to the south and Penglai to the north, passes through Zhaoyuan, and finally enters the Bohai Sea in eastern of Penglai, with a total length of approximately 100 km. The northern and southern sections of the fault show different attitudes; the overall trend is NE (30°–40°), and local deviations range from east to west, with a gentle dip in the north and middle sections. The strike of the Potouqing fault zone in the northern segment deflects eastward 50–70°. The southern section becomes steeper with a general orientation of 45°–60°. The faults generally develop along the boundary between the granite and Precambrian metamorphic basement and show characteristics of multi-phase activities, with ductile shear dominating in the early stage and brittle deformation in the late period (Lin, 2000; Gao et al. 2007).

The Muping–Jimo fault zone—located in the eastern part of the Jiaolai Basin—passes through Yantai Laishan, Taocun, and Haiyang Guocheng from north to south and connects with the Wulian-Rongcheng fault. The overall strike is NE (40°–50°), and the fault zone is approximately 300-km long and 20–60 km wide and includes four primary faults—from west to east: Taocun–Dongdoushan, Guocheng, Zhuwu–Yazi, and Yuli–Haiyang.

The Muping–Rushan fault zone is located in the northern Sulu UHP metamorphic belt, starting from the Yantai Muping to the north and Rushan to the south. The fault generally has NEE strikes (10°–20°), or NW inclines, and dips of 60–85° and primarily consists of five parallel and nearly equidistant faults from west to east: Qinghushan–Tangjiagou, Shigou Wushan, Sanjia, Jiangjunshi–Quhezhuang, and Gekou. These fracture zones are filled with gold-containing quartz–polymetallic sulphide veins.

The Mishan fault zone is located in Xishanhou-Hukouyao, Wendeng. The overall trend of the fault zone is south–north (SN), with an E incline, dip angle of 58°–76°, and length of approximately 30 km. The interior of the fault zone is primarily filled with granitic cataclastic rocks, the central part is filled with sericite and pyrite sericite, and quartz veins are widely distributed locally. The northern section of the fault zone is spread across the boundary between the Linglong and Weideshan plutons, and the southern section penetrates the Qierongcheng pluton.

2.2.3 Intrusions Rocks

The Mesozoic intrusions are divided into the following four suites: the Late Triassic Jiazishan Syenite (209–215 Ma) (Chen et al. 2003; Guo et al. 2005), the Late Jurassic Linglong biotite granite and the Luanjiahe monzogranite plutons (150–160 Ma) (Hou et al. 2007; Yang et al. 2012), the Cretaceous Guojialing granodiorite pluton (126–130 Ma) (Yang et al. 2012), and the latest-Cretaceous monzongranite plutons (e.g. Gushan and Aishan monzogranites) (110–120 Ma, Goss et al. 2010; Li et al. 2012; Charles et al. 2013).

(1) **Late Triassic intrusive rocks**

The Late Triassic intrusions are primarily distributed in the Shidao area in the eastern Sulu UHP metamorphic belt, including the Xiaoshidao, Jiazishan, Xingjia, and Cushan intrusions. Zircon U–Pb age dating shows that the formation is approximately 227–205 Ma (Chen et al. 2003; Guo et al. 2005). The principal lithologies are quartz syenite, pyroxene syenite, and alkaline gabbro.

(2) **Late Jurassic intrusive rocks**

The Late Jurassic intrusive rocks primarily include the Linglong, Luanjiahe, and Kunyushan plutons. The Linglong and Luanjiahe plutons primarily include biotite granite and biotite monzogranite. The Kunyushan pluton is primarily composed of quartz diorite, granodiorite, and monzodiorite. The Linglong granite was formed between 160 and 158 Ma, whereas the Luanjiahe granite was formed between 160 and 156 Ma (zircon U–Pb age; Miao et al. 1998; Yang et al. 2012; Ma et al. 2014). is distributed in the NE and EW directions, has an irregular ring-form with round and elliptical rock foundations; the pluton is controlled by the EW-direction basement fold structure of the Yishu fault and distributed with a "chang" shape into Laizhou, Zhaoyuan, Pingdu, Penglai, and other locations between the Zhao–Ping and Xincheng–Jiaojia fault zones. The Linglong pluton generally features a gradual transitional contact relationship with the surrounding Jiaodong and Fenzishan Groups. Linglong granites have medium-coarse, iso-grained, porphyritic structures, and massive structures and typically undergo partial metamorphism that produce long edges showing a gneiss-like structure. Most of the gold deposits in the Jiaodong area were produced in the Linglong granite or in the contact zone with metamorphic crystal basements.

(3) **Early Cretaceous intrusive rocks**

The Early Cretaceous intrusive rocks were dominated by Guojialing granite, which was formed at approximately 135–126 Ma (Zircon U–Pb dating; Wang et al. 1998; Yang et al. 2012) and a porphyritic hornblende-biotite granite primarily distributed in the Jiaobei terrane, which intrudes into the Linglong granite and metamorphic crystal basement (the contact zone) and is exposed in the Jiaodong Peninsula for approximately 500 km^2.

(4) **Latest Early Cretaceous intrusive rocks**

The Latest Early Cretaceous intrusions include the Gushan, Aishan, and Yashan plutons in the Jiaobei terrane and the Weideshan, Sanfoshan, Haiyang, and Laoshan plutons in the Sulu metamorphic belt. The lithology primarily consists of alkaline and monzonite granites (Goss et al. 2010; Li et al. 2012b). Zircon U–Pb chronology shows that the formation age of the granites ranges from 110–120 Ma (Goss et al. 2010; Li et al. 2012b; Charles et al. 2013).

2.2.4 *Dyke Swarms*

Large scale of dyke swarms emplaced in Jiaodong peninsula in late Mesozoic. These dykes were mostly lamprophyre, dolerite and diorite. They intruded at 85–135 Ma with a peak at ~120 Ma (Liu et al. 2009; Zhang et al. 2012; Cai et al. 2013, 2015; Ma et al. 2014; Deng et al. 2017). The occurrence of dyke swarms in the Jiaodong Peninsula is basically the same as that of regional fault structures, and is controlled by the regional geological structure. In gold mining areas, the distribution of dyke swarms are closely related to mineralized zones and fault zones. In the Zhao-Ping gold belt, the dyke swarms are mainly distributed in the directions of NNE, NNW and SN. For example, some ore bodies in the Linglong orefield are produced in lamprophyre veins, and some orebodies are strictly distributed on both sides of lamprophyre (Wang et al. 2008). In the Guocheng gold mining area, the dyke swarms are mostly NE40° ~50° trending, with a main tendency of about 70° and a width of 1–40 m (Tan et al. 2013). In the Muping-Rushan gold belt, the dyke swarms are mostly NE, NNE and NS directions, such as the Sanjia, Yazi, Pengjiakuang and Mouping gold mining areas (Sun et al. 2000; Guo et al. 2004). Regionally, the dykes in the Jiaobei terrane spread with direction of NNE and NE, with a small amount of NW, and the dykes in the Sulu orogenic belt distribute with the EW, ENE and NNE direction, with a small amount of NW direction (Fig. 2.3).

2.2.5 *Ore Resources*

The Jiaodong gold province is one of the largest gold-producing areas in China. The 159 gold mines in the Jiaodong Peninsula produce a combined 30 t of Au per year (Yang et al. 2014, 2016). Linglong-type and Jiaojia-type are the two principal types of gold deposits in the Jiaodong Peninsula. Linglong-type gold deposits indicate that gold-bearing quartz veins are formed in secondary faults that intrude into Mesozoic granites. Jiaojia-type gold mineralisation is largely disseminated and reticulated and located along first-level regional fractures and fracture-alteration zones, such as the Sanshandao and Jiaojia gold deposits (Deng et al. 2003; Li et al. 2015; Song et al. 2015). The gold deposits in the Jiaodong region are primarily controlled by regional NNE-trending subparallel faults, such as the Jiaojia, Sanshandao, and Zhaoping faults that cut through the Linglong and Guojialing granite plutons. Recent drilling results show that some of the gold deposits in the region, such as the Jiaojia and Sanshandao deposits, extend down to >2000 m from the surface (Hu et al. 2013). In addition to gold minerals, small amounts of silver, molybdenum, and iron ore were produced in the Jiaodong Peninsula.

References

Cai, Y.C., H.R. Fan, M. Santosh, F.F. Hu, K.F. Yang, and Z. Hu. 2015. Subduction-related metasomatism of the lithospheric mantle beneath the southeastern north china craton: Evidence from mafic to intermediate dykes in the northern sulu orogen. *Tectonophysics* 659: 137–151.

Cai, Y.C., H.R. Fan, M. Santosh, X. Liu, F.F. Hu, K.F. Yang, T.G. Lan, Y.H. Yang, and Y.S. Liu. 2013. Evolution of the lithospheric mantle beneath the southeastern North China Craton: Constraints from mafic dykes in the Jiaobei terrain. *Gondwana Research* 24: 601–621.

Charles, N., R. Augier, C. Gumiaux, P. Monié, Y. Chen, M. Faure, and R. Zhu. 2013. Timing, duration and role of magmatism in wide rift systems: Insights from the Jiaodong Peninsula (China, East Asia). *Gondwana Research* 24: 412–428.

Chen, L., C. Cheng, and Z. Wei. 2009. Seismic evidence for significant lateral variations in lithospheric thickness beneath the central and western North China Craton. *Earth Planetary Science Letters* 286, 171–183.

Chen, G.Y., D.S. Sun, X.R. Zhou, W. Shao, R.T. Gong, and Y. Shao. 1993. Mineralogy of Guojialing Granodiorite and Its Relationship to Gold Mineralization in the Jiaodong Peninsula. Beijing: China University of Geosciences press, 1–230 (in Chinese with English abstract)

Chen, J.F., Z. Xie, H.M. Li, X.D. Zhang, T.X. Zhou, Y.S. Park, K.S. Ahan, D.G. Chen, and X. Zhang. 2003. U-Pb zircon ages for a collision-related K-rich complex at Shidao in the Sulu ultrahigh pressure terrane, China. *Geochemical Journal* 37: 35–46.

Deng, J., W. Liu, Z.S. Sun, J.P. Wang, Q.F. Wang, Q.X. Zhang, Y.G. Wei. 2003. Evidence of mantle-rooted fluids and multi-level circulation ore-forming dynamics: A case study from the Xiadian gold deposit, Shandong Province, China. *Science in China (Series D: Earth Sciences)* 32: 123–134.

Deng, J., X.F. Liu, Q.F. Wang, and R.G. Pan. 2015. Origin of the Jiaodong-type Xinli gold deposit, Jiaodong Peninsula, China: Constraints from fluid inclusion and C-D-O-S-Sr isotope compositions. *Ore Geology Reviews* 65: 674–686.

Deng, J., X.F. Liu, Q.F. Wang, Y. Dilek, and Y.Y. Liang. 2017. Isotopic Characterization and Petrogenetic Modeling of the Early Cretaceous Mafic Diking-Lithospheric Extension in the North China Craton, Eastern Asia. *Geological Society of America Bulletin* 129: 1379–1407.

Dong, Y., J. Genser, and F. Neubauer. 2011. U-Pb and $^{40}Ar/^{39}Ar$ geochronological constraints on the exhumation history of the North Qinling terrane. *China. Gondwana Research* 19 (4): 881–893.

Goss, S.C., S.A. Wilde, F.Y. Wu, and J.H. Yang. 2010. The age, isotopic signature and significance of the youngest Mesozoic granitoids in the Jiaodong Terrane, Shandong Province, North China Craton. *Lithos* 120: 309–326.

Guo, J.H., F.K. Chen, X.M. Zhang, et al. 2005. Evolution of syn- to post-collisional magmatism from north Sulu UHP belt, eastern China: Zircon U-Pb geochronology. *Acta Petrologica Sinica* 21: 1281–1301 (in Chinese with English abstract).

Hou, M.L., Y.H. Jiang, S.Y. Jiang, H.F. Ling, and K.D. Zhao. 2007. Contrasting origins of late Mesozoic adakitic granitoids from the northwestern Jiaodong Peninsula, east China: Implications for crustal thickening to delamination. *Geology Management* 144: 619–631.

Hu, F.F., H.R. Fan, X.H. Jiang, X.C. Li, K.F. Yang, and T. Mernagh. 2013. Fluid inclusions at different depths in the Sanshandao gold deposit, Jiaodong Peninsula, China. *Geofluids* 13: 528–541.

Jiang, Y.H., S.Y. Jiang, H.F. Ling, and P. Ni. 2010. Petrogenesis and tectonic implications of Late Jurassic shoshonitic lamprophyre dykes from the Liaodong Peninsula, NE China. *Mineralogy and Petrology* 100: 127–151.

Li, J.W., S.J. Bi, and D. Selby, et al. 2012. Giant Mesozoic gold provinces related to the destruction of the North China craton. *Earth and Planetary Science Letters*, 349–350 (4): 26–37

Li, L., M. Santosh, and S.R. Li. 2015. The'Jiaodong type' gold deposits: Characteristics, origin and prospecting. *Ore Geology Reviews*, 589–611.

Li, X.H., Z.X. Li, W.X. Li, Y. Liu, Y. Chao, G.J Wei, and C.S. Qi. 2007. U–Pb zircon, geochemical and Sr–Nd–Hf isotopic constraints on age and origin of Jurassic I- and A-type granites from central Guangdong, SE China: A major igneous event in response to foundering of a subducted flat-slab? *Lithos*, 186–204.

Li, Z.L., and M.Z. Yang. 1993. The geochemistry of gold deposit in the Jiaodong Peninsula. Tianjing: Tianjin science and technology press, 1–301 (in Chinese with English abstract).

Liu, S., R. Hu, S. Gao, C. Feng, B. Yu, G. Feng, Y. Qi, T. Wang, and I.M. Coulson. 2009. Petrogenesis of Late Mesozoic mafic dykes in the Jiaodong Peninsula, eastern North China Craton and implications for the foundering of lower crust. *Lithos* 113: 621–639.

Ma, L., S.Y. Jiang, and A. Hofman. 2014. Lithospheric and asthenospheric sources of lamprophyres in the Jiaodong Peninsula: A consequence of rapid lithospheric thinning beneath the North China Craton? *Geochimica Et Cosmochimica Acta* 124: 250–271.

Mao, J.W., G.Q. Xie, Z.H. Zhang, X.F. Li, Y.T. Wang, C.Q. Zhang, and Y.F. Li. 2005. Mesozoic large-scale metallogenic pulses in north china and corresponding geodynamic setting. *Acta Petrologica Sinica* 21(1): 169–188 (In Chinese with English Abstract).

Song, M.C., S.Z. Li, M. Santosh, S. Zhao, S. Yu, P.H. Yi, S.X. Cui, G.X. Lv, J.X. Xu, Y.X. Song, and M.L. Zhou. 2015. Types, characteristics and metallogenesis of gold deposits in the Jiaodong Peninsula, Eastern North China Craton. *Ore Geology Reviews* 65: 612–625.

Tang, H.Y., J.P. Zheng, and C.M. Yu. 2008. Age and composition of the Rushan intrusive complex in the northern Sulu orogen, eastern China: petrogenesis and lithospheric mantle evolution. *Geology Magazine* 1: 1–17.

Wang, K.Y., C.Y. Zhang, and Y.M. Fan. 2008. Geochemical characteristics of ore-forming fluids of the Linglong gold deposit in Shandong Province[J]. *Journal of Jilin University (Earth Science Edition)* 38 (2): 194–201 (in Chinese with English abstract).

Wang, L.G., Y.M. Qiu, N.J. McNaughton, D.I. Groves, Z.K. Luo, J.Z. Huang, L.C. Miao, and Y.K. Liu, Y.K. 1998. Constraints on crustal evolution and gold metallogeny in the Northwestern Jiaodong Peninsula, China, from SHRIMP U–Pb zircon studies of granitoids. *Ore Geology Reviews*, 275–291.

Yang, J.H., F.Y. Wu, S.L. Chung, C.H. Lo, S.A. Wilde, and G.A. Davis. 2007. Rapid exhumation and cooling of the Liaonan metamorphic core complex: Inferences from 40Ar/39Ar thermochronology and implications for Late Mesozoic extension in the eastern North China Craton. *Geological Society of America Bulletin* 119: 1405–1414.

Yang, K.F., H.R. Fan, M. Santosh, F.F. Hu, S.A. Wilde, and T.G. Lan. 2012. Reactivation of the Archean lower crust: Implications for zircon geochronology, elemental and Sr–Nd–Hf isotopic geochemistry of Late Mesozoic granitoids from northwestern Jiaodong terrane, the North China Craton. *Lithos* 146–147: 112–127.

Yang, L.Q., J. Deng, L.N. Guo, Z.L. Wang, X.Z. Li, and J.L. Li. 2016. Origin and evolution of ore fluid, and gold-deposition processes at the giant Taishang gold deposit, Jiaodong Peninsula, eastern China. *Ore Geology Reviews* 72: 585–602.

Yang, L.Q., J. Deng, Z.L. Wang, and L. Zhang. 2014. Mesozoic gold metallogenic system of the Jiaodong gold province, eastern China. *Acta Geologica Sinica* 30 (9): 2447–2467 (in Chinese with English abstract).

Yang, M.Z., and G.X. Lu. 1996. The geology and geochemistry of gold deposit in the greenstone belt of east Shangdong province. Beijing: Geology press, 1–228 (in Chinese with English abstract).

Yang, Z.F. 1998. Crustal evolution and gold mineralization geochemistry in Jiaodong region. Geological Publishing House, Beijing (In Chinese with English Abstract).

Zhai, M.G. 2010. Tectonic evolution and metallogenesis of North China Craton. *Mineral Deposits* 29 (1): 24–31 (In Chinese with English abstract).

Zhai, M.G., and A.G. Bian. 2000. The Neoarchean Super-Continent cohere and Late Paleoproterozoic- Mesoproterozoic disintegration of the North China Craton. *Science in China (Series D)* 30 (S1): 129–137 (in Chinese with English abstract).

Zhai, M.G., H.R. Fan, J.H. Yang, and L.C. Miao. 2004. Large-scale cluster of gold deposits in eastern Shandong: Anorogenic metallogenesis. *Geoscience Frontiers* 11: 85–98 (in Chinese with English abstract).

Zhai, M.G., and M. Santosh. 2013. Metallogeny of the North China Craton: Link with secular changes in the evolving Earth. *Gondwana Research* 24: 275–297.

Zhang, J., Z.F. Zhao, F.Y. Zheng, X.M. Liu, and L.W. Xie. 2012. Zircon Hf–O isotope and whole-rock geochemical constraints on origin of postcollisional mafic to felsic dykes in the Sulu orogen. *Lithos*, 136–139, p. 225–245.

Zhang, L.C., T.B. Liu, Y.C. Shen, Q.D. Zeng, and G.M. Li. 2003. Structure, Isotopes, and 40Ar/39Ar Dating of the Pengjiakuang Gold Deposit, Mesozoic Jiaolai Basin, Eastern China. *International Geology Review*: 691-711.

Zhao, G.C., 2001. Palaeoproterozoic assembly of the North China Craton. *Geological Magazine* 138: 87–91.

Zhao, G.C., M. Sun, A.W. Simon, and S. Li. 2005. Late Archean to Paleoproterozoic evolution of the North China Craton: Key issues revisited. *Precambrian Research*136: 177–202.

Zhao, G.C., and M.G. Zhai. 2013. Lithotectonic elements of Precambrian basement in the North China Craton: Review and tectonic implications. *Gondwana Research* 23: 1207–1240.

Zhou, X.M., and W.X. Li. 2000. Origin of Late Mesozoic igneous rocks in Southeastern China: Implications for lithosphere subduction and underplating of mafic magmas: Tectonophysics, 326: 269–287.

Zhu, G., M.L. Niu, C.L. Xie, and Y. Wang. 2010. Sinistral to normal faulting along the Tan-Lu fault zone: Evidence for geodynamic switching of the East China continental margin. *the Journal of Geology* 118 (3): 277–293.

Chapter 3
Micro-Geochemistry of Clinopyroxene of the Mafic Dykes in the Jiaodong Peninsula: Implications for Petrogenesis and Geodynamic

Abstract Early Cretaceous mafic dykes widely spread in the Jiaodong Peninsula and provided significant evidence for geodynamic of the studied area. Based on the whole geochemistry, in situ major element, trace element, Sr isotopic compositions and H_2O contents analyses of the clinopyroxene phenocrysts of the mafic dykes were analysed. These evidences proved that the lithospheric mantle beneath Jiaodong Peninsula had undergone the metasomatism resulted from Paleo-Pacific slab subduction, and that lithospheric thinning of the eastern NCC caused by prolonged thermo-mechanical erosion. Both of these processes played a significant role in the petrogenesis of the Mesozoic mafic dykes at Jiaodong.

Keywords Mafic dykes · Clinopyroxene phenocrysts · Lithospheric mantle · Metasomatism · Lithospheric thinning

3.1 Introduction

Mafic dykes, which are common in extensional tectonic settings, provide insights into the composition of the mantle and the regional geodynamic setting (Halls 1982; Ernsrt et al. 1995; Sun et al. 2000; Hou et al. 2008; Deng et al. 2014, 2015; Wang et al. 2014, 2016c). Numerous studies have been conducted on the petrology, mineralogy, geochemistry, and geochronology of mafic dykes worldwide, to determine their petrogenesis and to provide further information regarding their tectonic evolution (Liu et al. 2009, 2013a, b; Callegaro et al. 2013; Krüger et al. 2013; Batki et al. 2014; Deng et al. 2014, 2015c; Wang et al. 2016a, b). Mesozoic mafic dykes are widespread on the Jiaodong Peninsula (including Jiaobei terrain and Sulu orogenic belt), located in the southeastern part of the North China Craton (NCC) (Guo et al. 2004; Liu et al. 2006, 2008b, 2009; Deng et al. 2015a; Tan et al. 2015). The tectonic significance of these dykes and the intimate relationship between the mafic dykes and gold lodes have been the topics of extensive research (Guo et al. 2004, 2014; Yang et al. 2004; Liu et al. 2008a, 2012, 2015; Tan et al. 2007, 2013; Cai et al. 2013, 2015; Ma et al. 2014, 2016). Numerous studies have been conducted on the geochronology (e.g., zircon U–Pb dating), geochemistry (e.g., whole-rock major,

Y. Liang, *Mineral Micro-Geochemistry Constraints on Petrogenesis and Genesis of Gold Deposit*, https://doi.org/10.1007/978-981-16-1022-6_3

trace and rare earth elements) and isotopic compositions (e.g., whole-rock Sr–Nd–Pb isotopes) of the mafic dykes at Jiaodong in order to determine their petrogenesis and provide further information on the tectonic evolution of the area (Liu et al. 2009; Ma et al. 2014, 2016; Cai et al. 2015; Deng et al. 2017). A series of whole-rock geochemical data for the lithospheric mantle-derived dykes is available; however, no consensus has been reached on the geodynamic setting of their formation (Yang et al. 2004; Guo et al. 2004; Liu et al. 2009; Tan et al. 2013; Ma et al. 2014; Cai et al. 2015). Moreover, the Jiaodong mafic dykes display similar Sr–Nd isotopic compositions to those of the recycled lower crust of the NCC and Yangtze Craton (YC) as well as seawater-altered oceanic crust with marine sediments, which further hinders identification of the process of mantle source enrichment in this region. The petrogenesis and tectonic dynamics background of these dykes remain controversial. To reconstruct the composition of the mantle beneath Jiaodong Peninsula and the regional geodynamic setting, it is necessary to have a better understanding of the mafic dyke magmatism in this region (Rock 1987; Ernst et al. 1995; Qiu and Deng 2016; Wang et al. 2016; Tahmasbi et al. 2016). Clinopyroxenes, which form as early-stage crystals during magma cooling and evolution, preserve better records of the primary features of the magma than whole-rock geochemical data, and for this reason we have studied the clinopyroxene phenocrysts in the Jiaodong mafic dykes to obtain new insights into the evolution of the magma and the nature of the source material. This chapter chose Early Cretaceous mafic dykes in the southeastern Jiaolai basin (near Haiyang city), Linglong gold deposit, Laiyang area, Muping area and Wanggezhuang area, Jiaodong Peninsula (Fig. 3.1), as the studied objects. In this chapter, we report on the detailed petrography and geochemistry of the clinopyroxene phenocrysts, including in situ determinations of major element, trace element, Sr isotopic compositions and H_2O contents. We combine these new data with previous whole-rock geochemical data and assess the extent of metasomatism in the source rocks and magma mixing prior to eruption. We provide further evidence for metasomatism of the lithospheric mantle beneath Jiaodong Peninsula as a result of Paleo-Pacific slab subduction, and for thinning of the lithospheric mantle below the NCC as a result of continuous thermo-mechanical erosion.

3.2 Petrology of the Early Cretaceous Mafic Dykes in the Jiaodong Peninsula

Mafic dykes in the Jiaodong Peninsula mainly were composed of dolerite and lamprophyre. These dykes were intruded into Precambrian basement of granites (striking NE and NNE) and were 0.2–3.0 m wide (Figs. 3.2 and 3.3). The lamprophyre dykes have been divided into spessartite and odinite (according to the scheme of Woolley et al. 1996), and show porphyritic structures. The phenocrysts of the odinite are mainly clinopyroxene (5–15%). Its groundmass includes microphenocrysts of amphibole (5%–10%), biotite (10%–15%), plagioclase (20–30%), alkali feldspar (5%–10%),

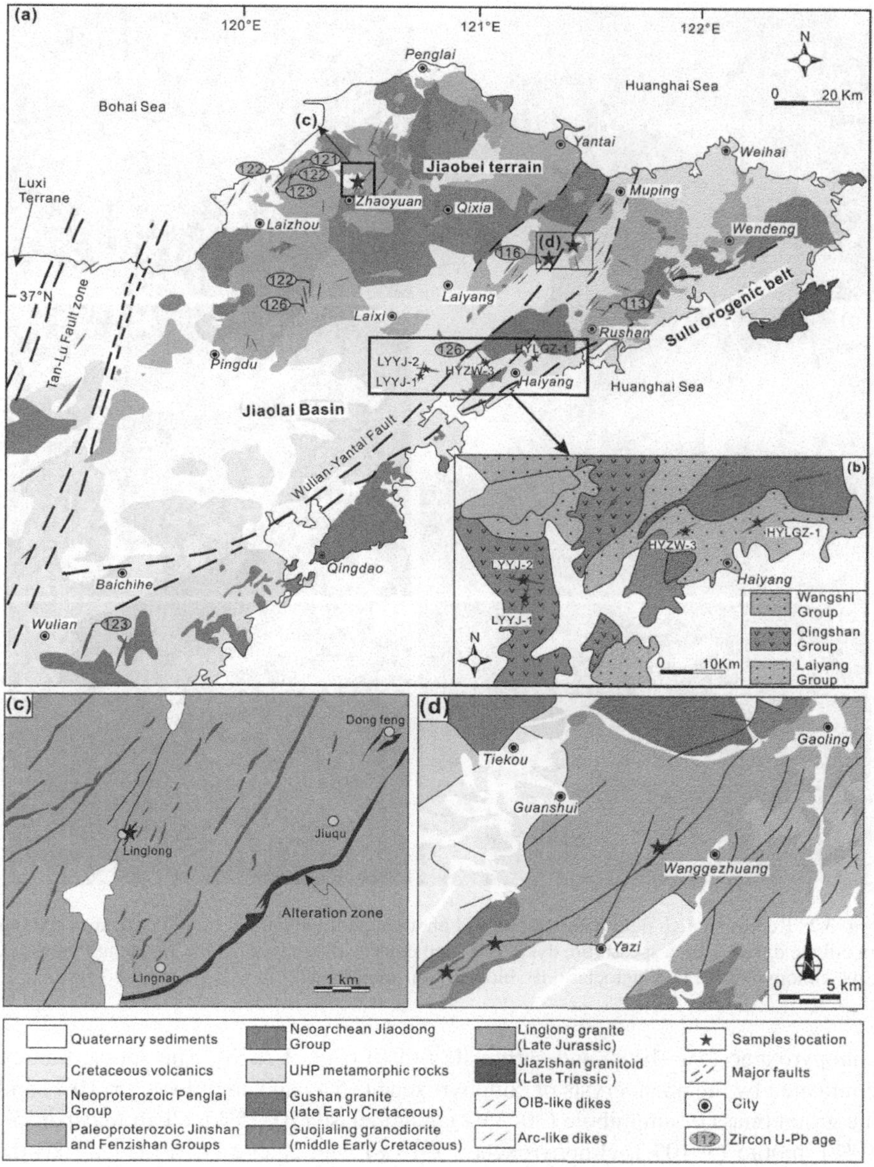

Fig. 3.1 **a** Simplified geological map of the Jiaodong Peninsula (modified after Cai et al. 2015), showing the distribution of mafic dykes (mafic dykes are after Liu et al. 2009; Cai et al. 2013; Guo et al. 2014; Ma et al. 2014, 2016), **b** southeastern Jiaolai basin (modified after Xie et al. 2012), **c** the Linglong gold field and **d** Laiyang and Muping area with location of the mafic dykes

Fig. 3.2 Representative field photographs and photomicrographs of lamprophyre dykes showing an odinite dyke (**a**, **b**), a spessartite dyke (**c**, **d**) and dolerite dykes (**e**, **f**) in the Jiaodong Peninsula. Cpx, clinopyroxene; Pl, plagioclase; Bt, biotite; Am, amphibole

clinopyroxene (5%–10%) and carbonate (~5%) (Fig. 3.2a–b). The spessartite are dominated by the phenocrysts of clinopyroxene (~5%) and amphibole (5–10%) and the groundmass of amphibole (30–35%), plagioclase (20–25%), alkali feldspar (5–10%), biotite (5–10%), clinopyroxene (5–10%), and calcite (~5%) (Fig. 3.2c–d). The dolerite dykes are composed mainly of clinopyroxene (30–40%) and plagioclase (50–60%), with minor amphibole (~5%) and biotite (~5%) (Fig. 3.2e–f).

Although most of the phenocrysts in these dykes have been heavily altered, the clinopyroxene phenocrysts are fresh, and they fall into two groups. Group I clinopyroxenes show clear zoning with core–rim structures (up to 3 mm; Fig. 3.3a), whereas Group II clinopyroxenes are subhedral and lack any clear zoning (up to 2 mm; Fig. 3.3b).

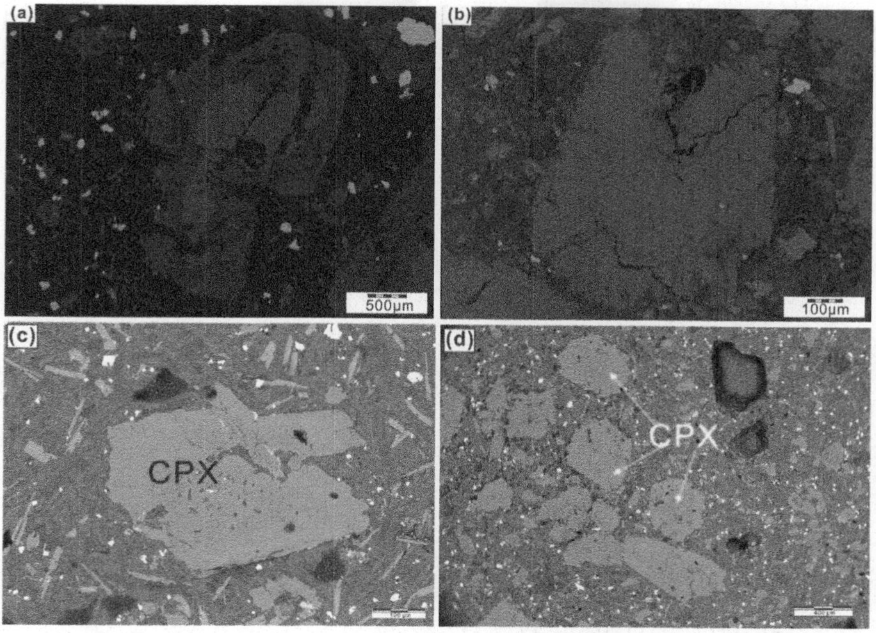

Fig. 3.3 The representative photomicrographs and Back-scattered electron (BSE) images of Jiaodong mafic dyke with Cpx phenocrysts. **a** Group I clinopyroxenes showing clear zoning with core–rim structures; **b** Group II clinopyroxenes with lack of any clear zoning; **b** High-Mg clinopyroxenes of the mafic dykes in Laiyang area

3.3 Emplacing Age of the Early Cretaceous Jiaodong Mafic Dykes

The geochronology, including LA-ICP-MS or SHRIMP zircon U-Pb dating, has implied that Laiyang mafic dykes age of 125.6 ± 4.3 Ma (Liang et al. 2017), Wanggezhuan mafic dykes age of 114–116 Ma (Liu et al. 2020). Two mafic dykes of a lamprophyre (MPWGZ-1) and a dolerite (LYGC-2) sampled on Muping area, Jiaodong Peninsula, emplaced at 119 ± 3 Ma (Deng et al. 2017). Detailed geochronological research on the mafic dyke swarms in the Jiaodong Peninsula have been conducted, which imply that the mafic dyke swarms were emplaced at 110–130 Ma (Table 3.1).

Table 3.1 The age from intermediate-mafic dykes in the Jiaodong Peninsula (After Liu et al. 2020)

Location	Type	Age/Ma	Method	Mineral	References
Jiaobei-Jiaojia	Diabase-porphyrite	122.4 ± 3.1	LA-ICP-MS U-Pb	Zircon	Ma et al. (2014)
Jiaobei-Jiaojia	Diabase-porphyrite	122.6 ± 3.3	LA-ICP-MS U-Pb	Zircon	
Jiaobei-Jiaojia	Lamprophyre	120.8 ± 1.8	LA-ICP-MS U-Pb	Zircon	
Jiaobei-Laixi	Diabase-porphyrite	123.7 ± 2.2	LA-ICP-MS U-Pb	Zircon	Liu et al. (2009)
Jiaobei-Laixi	Diabase-porphyrite	125.4 ± 1.1	LA-ICP-MS U-Pb	Zircon	
Jiaobei-Laixi	Diabase-porphyrite	122.5 ± 1.5	LA-ICP-MS U-Pb	Zircon	
Jiaobei-Laixi	Diabase-porphyrite	126.9 ± 1.7	LA-ICP-MS U-Pb	Zircon	
Jiaobei-Laixi	Diabase-porphyrite	124.8 ± 1.5	LA-ICP-MS U-Pb	Zircon	
Sulu-Rushan	Lamprophyre	113 ± 1	LA-ICP-MS U-Pb	Zircon	Cai et al. (2015)
Sulu-Junan	Diabase	119 ± 1.7	LA-ICP-MS U-Pb	Zircon	Liu et al. (2008)
Sulu-Junan	Diabase	120.2 ± 1.9	LA-ICP-MS U-Pb	Zircon	
Jiaobei	Lamprophyre	120.7 ± 2.9	LA-ICP-MS U-Pb	Zircon	Deng et al. (2017)
Jiaobei	Lamprophyre	120.1 ± 1.3	LA-ICP-MS U-Pb	Zircon	
Jiaobei	Lamprophyre	118.7 ± 2.2	LA-ICP-MS U-Pb	Zircon	
Jiaobei	Lamprophyre	111.7 ± 1.2	LA-ICP-MS U-Pb	Zircon	
Sulu	Lamprophyre	130.1 ± 3.9	LA-ICP-MS U-Pb	Zircon	
Sulu	Lamprophyre	118.5 ± 2.9	LA-ICP-MS U-Pb	Zircon	
Sulu	Lamprophyre	119.3 ± 0.8	LA-ICP-MS U-Pb	Zircon	

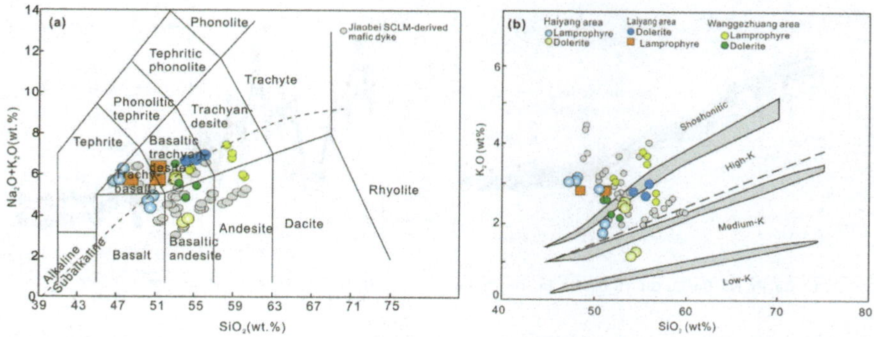

Fig. 3.4 **a** $Na_2O + K_2O$ versus SiO_2 and **b** K_2O versus SiO_2 plots for the Jiaodong mafic dykes. *Data source* Jiaobei SCLM-derived maifc dykes are from Deng et al. (2017); Mafic dykes in the Haiyang and Laiyang are from Liang et al. (2017 and 2019); Mafic dykes in the Wangezhuang area are from Liu et al. (2020)

3.4 Petrogenesis of the Mafic Dykes

3.4.1 Geochemical Compositions of the Jiaodong Mafic Dykes

The dolerite and lamprophyre dykes in the Jiaodong Peninsula share similar geochemical compositions. Jiaodong mafic dykes exhibited SiO_2 contents of 45–53 wt% and $Na_2O + K_2O$ contents of 2.1–6.7 wt%, belonging to basalt and andesite + basaltic andesite (Fig. 3.4a). The dykes display K_2O contents of 1.1–2.9 wt%, which was classified as calc-alkaline to high-K calc-alkaline (Fig. 3.4b). These mafic dykes show high contents of MgO (6.8–10.13 wt%), Cr (267–702 ppm), and Ni (114–335 ppm), and Mg# value of 65–72. The mafic dykes show significantly fractionated REE patterns and lack of Eu anomalies (Fig. 3.5a). They are enriched in large ion lithophile elements (LILEs) and depleted in high field strength elements (HFSEs) (Fig. 3.5b). These characteristics are similar to those of Aleutian arc basalts (Kelemen et al. 2003) and lithospheric mantle-derived mafic dykes (110–130 Ma) on the Jiaodong Peninsula (Ma et al. 2014). Jiaodong mafic dykes show $(^{87}Sr/^{86}Sr)_i$ isotope ratios of 0.7075–0.7081) and $\varepsilon_{Nd}(t)$ values of (−18 to −13) (Fig. 3.6) (Tables 3.2 and 3.3).

3.4.2 Limited Crustal Contamination

Before discussing the genesis of dyke rocks at depth, we first make necessary assessments of whether these dyke rocks have undergone low-temperature alteration and crustal assimilation and contamination. Although a few diabase and lamprophyre

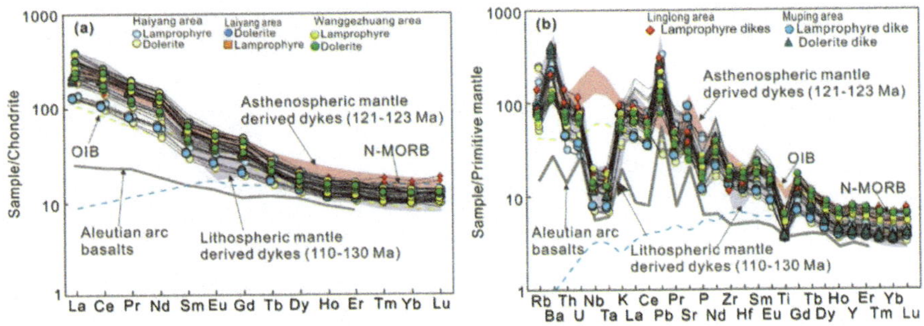

Fig. 3.5 **a** Chondrite-normalized REE patterns and **b** primitive mantle-normalized immobile trace element patterns of the Jiaodong mafic dykes. Data sources: Asthenospheric mantle-like mafic dykes (123–121 Ma) are from Ma et al. (2014, 2016); Lithospheric mantle-derived mafic dykes (130–110 Ma) are from Liu et al. (2009), Ma et al. (2014) and Deng et al. (2017); Mafic dykes in the Linglong, Muping, Haiyang and Laiyang are from Liang et al. (2017, 2018, 2019 and 2020a); Mafic dykes in the Wangezhuang area are from Liu et al. (2020)

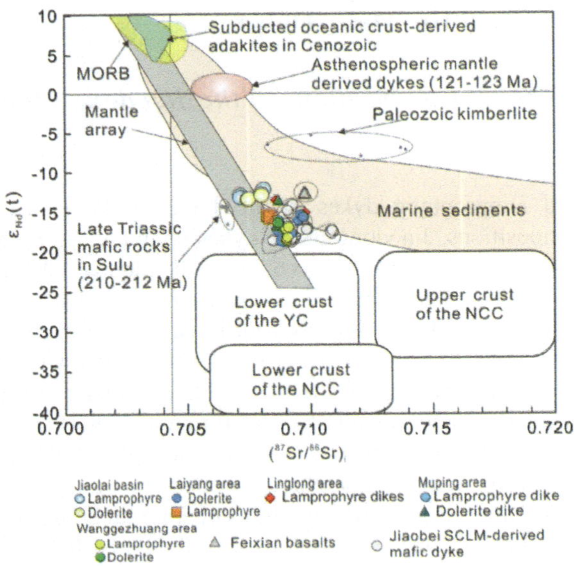

Fig. 3.6 ($^{87}Sr/^{86}Sr$)i versus $\varepsilon_{Nd}(t)$ plot for the Jiaodong mafic dykes. Data sources: Palaeozoic kimberlites and peridotites are from Zheng and Lu (1999); Triassic mafic dykes in Sulu orogenic belt are from Zhao et al. (2012); Cenozoic subducted oceanic crust-derived adakites are from Sajona et al. (2000); upper and lower NCC crust and lower YC crust are from Jahn et al. (1999); Asthenospheric mantle-derived dykes (~122 Ma) are from Ma et al. (2014, 2016), and Jiaobei SCLM-derived mafic dykes are from Deng et al. (2017), and Feixian basalt are from Gao et al. (2008); Mafic dykes in the Linglong, Muping, Haiyang and Laiyang are from Liang et al. (2017, 2018, 2019 and 2020a); Mafic dykes in the Wangezhuang area are from Liu et al. (2020)

Table 3.2 Geochemical compositions of the mafic dykes in Jiaodong Peninsula

Sample	LYYJ-2-1	LYYJ-2-2	LYYJ-2-3	LYYJ-1-1	LYYJ-1-2	HYZW-3-1	HYZW-3-2	HYZW-3-3	HYLGZ-1-1	HYLGZ-1-2	HYLGZ-1-3	LLW50-1	LLW50–19
	Haiyang mafic dykes (Data are from Liang et al. 2017)											Lilong mafic dykes	
SiO_2	52.3	52.3	52.2	50.73	51.53	47.98	48.39	48.63	44.26	44.87	45.31	45.31	50.27
TiO_2	1.21	1.21	1.2	0.8	0.81	1.11	1.13	1.18	1.17	1.11	1.12	0.72	0.8
AL_2O_3	16	15.99	16.09	13.98	14.21	14.85	15.24	15.72	14.49	14.94	14.71	12.09	13.93
TFe_2O_3	6.6	6.7	6.7	6.91	6.57	8.13	7.97	7.91	8.68	8.03	8.08	7.36	7.2
FeO	5.53	5.57	5.68	5.03	4.99	5.93	5.32	5.07	4.78	4.67	4.35	5.08	5.12
Fe_2O_3	1.19	1.26	1.13	2.09	1.76	2.44	2.94	3.16	4.33	3.73	4.15	1.72	1.51
MnO	0.06	0.06	0.06	0.09	0.07	0.18	0.15	0.15	0.15	0.14	0.15	0.13	0.13
MgO	7.32	7.28	7.35	10.13	9.8	9.38	8.49	7.55	9.09	7.95	8.25	11.05	9.2
CaO	8.12	8.14	8.14	7.21	7.4	7.92	8.65	8.2	9.13	8.87	9.97	8.08	8.53
Na_2O	3.3	3.36	3.32	2.4	2.48	1.81	2.52	2.86	2.55	2.93	2.59	1.66	2.38
K_2O	2.48	2.34	2.36	1.08	1.18	2.69	1.63	1.87	2.82	2.92	2.87	2.12	2.15
P_2O_5	0.42	0.41	0.41	0.24	0.24	0.47	0.48	0.53	0.92	0.91	0.86	0.22	0.29
LOI	1.11	1.04	1.03	5.22	4.61	4.13	4.08	4.03	5.38	5.92	4.69	12.43	0.9
Total	99.04	98.96	98.97	99	99.08	98.89	99.02	98.95	99.07	98.96	99.02	100.6	95.2
Mg#	66.41	65.94	66.18	72.32	72.66	67.3	65.52	62.97	65.13	63.84	64.53	75	72
La	42.5	44.8	45.5	30.5	29.8	31.9	32.4	32.6	65.9	67.8	64.2	37.6	47.3
Ce	83.2	87.8	88.8	59.5	58.8	64.6	67.2	67.5	139	137	133	67.6	87.5
Pr	9.26	9.86	9.87	6.53	6.39	7.46	7.76	7.71	15.7	15.4	15.5	8.02	10.19
Nd	32.3	35.3	37.2	24.3	22.1	28.2	30.2	28.6	59.3	60	62	30.6	37.4
Sm	6.14	6.32	6.43	4.55	4.26	5.76	5.93	5.82	10.8	10.1	10.6	5.14	6.14

(continued)

Table 3.2 (continued)

Sample	LYYJ-2-1	LYYJ-2-2	LYYJ-2-3	LYYJ-1-1	LYYJ-1-2	HYZW-3-1	HYZW-3-2	HYZW-3-3	HYLGZ-1-1	HYLGZ-1-2	HYLGZ-1-3	LLW50-1	LLW50–19
	Haiyang mafic dykes (Data are from Liang et al. 2017)											Lilong mafic dykes	
Eu	1.74	1.8	1.81	1.31	1.25	1.73	1.81	1.78	2.94	2.86	2.98	1.49	1.71
Gd	5.47	5.49	5.35	3.85	3.92	5.44	5.28	5.23	8.76	8.28	8.69	4.14	4.8
Tb	0.74	0.73	0.71	0.55	0.53	0.75	0.7	0.71	1.05	0.98	1.07	0.58	0.66
Dy	4.18	4.15	3.98	3.17	3.02	4.14	4.08	3.97	5.29	5.1	5.22	3.09	3.48
Ho	0.79	0.8	0.77	0.61	0.59	0.78	0.79	0.77	0.93	0.9	0.94	0.58	0.63
Er	2.24	2.18	2.2	1.74	1.6	2.12	2.19	2.18	2.51	2.37	2.36	1.51	1.65
Tm	0.33	0.34	0.33	0.27	0.26	0.33	0.34	0.33	0.37	0.35	0.36	0.25	0.28
Yb	1.96	2.01	2	1.57	1.56	1.94	1.97	1.92	2.17	2.02	2.16	1.53	1.67
Lu	0.3	0.31	0.31	0.24	0.24	0.29	0.3	0.29	0.33	0.3	0.32	0.26	0.29
Rb	159	153	155	32.9	36.3	108	67.3	79.6	60.1	63.6	61.1	71.9	40.6
Ba	931	885	817	817	831	1830	1493	1572	2046	2051	2133	1233	1204
Th	4.64	4.73	4.63	3.89	3.98	2.62	2.67	2.62	6.2	6.58	6.22	6.61	9.6
U	0.94	0.95	0.93	0.76	0.74	0.58	0.57	0.59	1.27	1.32	1.28	1.45	1.9
Nb	9.43	9.49	9.16	4.83	4.89	11.3	12.2	12.9	12.7	13.4	12.8	4.98	7.39
Ta	0.55	0.55	0.56	0.27	0.28	0.61	0.63	0.66	0.63	0.66	0.65	0.35	0.47
K	20587	19425	19591	8965	9796	22331	13531	15523	23410	24240	23825	17599	17848
Pb	2.02	2.09	1.91	8.28	7.22	23.6	10.8	13.7	10.4	12.4	10.3	19.2	13.9
Sr	865	862	848	699	687	629	738	766	1381	1366	1420	539	718
P	1833	1789	1789	1048	1048	2051	2095	2313	4015	3972	3754	945	1271

(continued)

Table 3.2 (continued)

Sample	LYYJ-2-1	LYYJ-2-2	LYYJ-2-3	LYYJ-1-1	LYYJ-1-2	HYZW-3-1	HYZW-3-2	HYZW-3-3	HYLGZ-1-1	HYLGZ-1-2	HYLGZ-1-3	LLW50-1	LLW50-19
	Haiyang mafic dykes (Data are from Liang et al. 2017)											Lilong mafic dykes	
Zr	193	192	186	137	138	188	203	204	221	230	220	72.4	100.6
Hf	4.46	4.42	4.45	3.23	3.29	4.75	4.6	4.63	5.25	4.99	5.15	2.44	3.32
Ti	7254	7254	7194	4796	4856	6654	6774	7074	7014	6654	6714	4293	4782
Y	20.4	21.8	20	16.1	15.5	19.9	20.7	20.8	26.9	26.9	27.3	14.9	16.4
V	162	162	160	162	158	206	208	209	202	195	199	168	176
Cr	267	271	279	652	617	466	387	266	278	196	294	702	513
Co	27.5	26.5	27.6	45.2	42.2	38.2	36.5	32.7	35.1	29.9	30.2	38.81	33.02
Ni	123	124	122	335	322	137	127	100	119	92.6	101	243.6	152.7

Sample	LY01-1	LY01-2	LY01-3	LY02-1	LY02-3	LY02-4	MP-1-1	MP-1-3	MP-1-5	LYGC-2-1	LYGC-2-3	LYGC-2-4
	Laiyang mafic dykes (Data are from Liang et al. 2018)						Muping mafic dykes (Data are from Deng et al. 2017)					
SiO_2	51.52	51.74	52.76	48.63	48.45	45.6	50.2	51.5	50.9	50.7		49.1
TiO_2	0.81	0.81	0.81	0.82	0.84	0.8	0.76	0.76	0.77	0.83		0.81
AL_2O_3	14.2	16.1	14.5	13.5	13.8	16.7	12	12.6	12.2	12.8		12.6
TFe_2O_3	6.64	6.75	6.7	7.85	8.1	7.62	1.21	1.62	1.63	2.96		2.84
MnO	0.09	0.09	0.09	0.11	0.12	0.11						
MgO	6.85	7.62	7.53	9.39	9.47	9.12	12	11.6	11.6	11.5		11.8
CaO	5.94	4.94	4.66	7.46	7.42	7.92	7.15	6.63	7.08	6.89		7.62
Na_2O	3.94	3.87	3.87	3.53	2.93	2.87	1.95	2.16	2.14	2.58		2.41
K_2O	2.48	2.66	2.84	2.68	2.64	2.67	1.49	1.75	1.55	2.42		2.51

(continued)

Table 3.2 (continued)

Sample	LY01-1	LY01-2	LY01-3	LY02-1	LY02-3	LY02-4	MP-1-1	MP-1-3	MP-1-5	LYGC-2-1	LYGC-2–3	LYGC-2–4
	Laiyang mafic dykes (Data are from Liang et al. 2018)						Muping mafic dykes (Data are from Deng et al. 2017)					
P_2O_5	0.42	0.43	0.42	0.45	0.45	0.43	0.26	0.27	0.26	0.52		0.5
LOI	6.48	5.3	5.31	5.86	5.32	5.68	6.87	5.53	6.34	3.99		5.18
Total	99.33	100.35	99.48	100.28	99.57	99.52	99.8	99.8	99.8	99.8		99.8
Mg#	65	67	67	68	68	68	73.9	73.4	73.4	72		73
La	66.2	68.3	69.4	64.6	69.2	63	30.4	30.5	29.6	53.2		48.9
Ce	125.2	123.8	127.3	125.4	131.9	121.9	64.9	64.5	62.5	124		124
Pr	13.5	13.3	13.7	14	14.7	13.5	7.81	7.68	7.49	14.9		14.5
Nd	50.2	49.7	50.5	53.5	55.2	51.5	30	29.6	28.2	56.8		55.7
Sm	7.69	7.32	7.59	8.36	8.69	8.25	5.1	5.01	4.89	9.08		8.64
Eu	1.93	2.01	2.08	2.21	2.36	2.24	1.48	1.55	1.45	2.49		2.44
Gd	5.35	5.09	5.25	5.8	6.12	5.73	4.29	4.2	4.07	6.81		6.62
Tb	0.66	0.64	0.69	0.75	0.76	0.73	0.63	0.62	0.6	0.87		0.86
Dy	3.65	3.52	3.62	3.96	4.28	3.93	3.32	3.27	3.24	4.16		3.98
Ho	0.64	0.64	0.64	0.73	0.74	0.72	0.63	0.61	0.6	0.72		0.68
Er	1.83	1.83	1.84	2.01	2.13	1.97	1.72	1.72	1.71	2.05		1.96
Tm	0.26	0.23	0.24	0.26	0.28	0.27	0.25	0.26	0.25	0.29		0.27
Yb	1.52	1.42	1.5	1.65	1.71	1.71	1.66	1.68	1.6	1.88		1.76
Lu	0.24	0.23	0.23	0.24	0.27	0.27	0.27	0.27	0.26	0.3		0.3
Hf	3.95	4.22	4.54	4.33	3.93	4.26	3.51	3.89	3.67	4.33		4.26

(continued)

Table 3.2 (continued)

Sample	LY01-1	LY01-2	LY01-3	LY02-1	LY02-3	LY02-4	MP-1-1	MP-1-3	MP-1-5	LYGC-2-1	LYGC-2–3	LYGC-2–4
	Laiyang mafic dykes (Data are from Liang et al. 2018)						Muping mafic dykes (Data are from Deng et al. 2017)					
Ta	0.46	0.44	0.44	0.38	0.41	0.36	0.31	0.31	0.32	0.42		0.42
Th	9.33	8.54	8.97	8.7	9.44	8.71	3.82	3.72	3.85	8.54		8.44
U	1.7	1.48	1.58	1.55	1.75	1.63	0.74	0.72	0.79	1.73		1.68
Nb	9.56	8.94	9.45	7.75	8.39	7.77	5.51	5.51	5.7	8.41		8.35
Rb	49.8	55.2	59.6	51.9	54	55.5						
Ba	2014	1892	2027	2623	3109	2150	1347	1449	1542	2324		2862
Pb	14.22	10.7	11.24	25.46	14.87	17.4						
Sr	921	1185	1374	683	975	828	1433	1925	1446	898		820
Zr	166.7	181.6	176.1	162.3	184.3	176.4	144	156	146	179		176
Y	18.8	24.6	17.8	20.3	21.7	19.8	17.6	17.3	16.9	21.1		19.8
V	134	134	136	141	152	144	169	168	171	166		171
Cr	299	309	312	445	473	436	584	557	557	417		397
Co	26.1	26.5	26.7	31.6	34	31						
Ni	114	121	120	184	201	176						

Sample	WGZ-4-1	WGZ-4-2	WGZ-4-3	WGZ-5-1	WGZ-5-2	WGZ-6-1	WGZ-6-2	WGZ-7-1	WGZ-7-2	WGZ-8-1	WGZ-8-2
Rock type	Lamprophyre dike			Lamprophyre dike		Dolerite dike		Lamprophyre dike		Dolerite dike	
	Wanggezhuang mafic dykes (Data are from Liu et al. 2020)										
SiO_2	53.96	53.82	54.26	55.4	55.17	51.5	51.77	51.37	51.85	50.92	50.49
TiO_2	0.73	0.7	0.7	0.66	0.67	0.83	0.84	0.99	1	0.87	0.88

(continued)

Table 3.2 (continued)

Sample	WGZ-4-1	WGZ-4-2	WGZ-4-3	WGZ-5-1	WGZ-5-2	WGZ-6-1	WGZ-6-2	WGZ-7-1	WGZ-7-2	WGZ-8-1	WGZ-8-2
Rock type	Lamprophyre dike			Lamprophyre dike		Dolerite dike		Lamprophyre dike		Dolerite dike	
	Wanggezhuang mafic dykes (Data are from Liu et al. 2020)										
Al_2O_3	14.92	14.89	14.91	14.73	14.73	15.39	15.79	15.18	15.43	14.68	14.57
TFe_2O_3	6.29	6.48	6.92	6.66	6.41	8.71	7.43	7.42	7.53	8.05	8.02
MnO	0.1	0.09	0.09	0.08	0.07	0.11	0.11	0.13	0.12	0.12	0.11
MgO	6.17	6.04	5.83	6.29	6.1	8.33	7.19	7.78	7.65	8.6	8.4
CaO	5.87	5.78	5.6	5.77	6.1	7.68	7.68	7.27	6.98	8.27	8.06
Na_2O	3.63	3.27	3.31	3.34	3.18	2.72	3.3	3.14	3.25	3.05	3.98
K_2O	3.65	3.31	3.53	2.43	2.62	2.16	2.05	3.02	2.96	2.52	2.51
P_2O_5	0.45	0.43	0.43	0.34	0.35	0.28	0.32	0.65	0.66	0.5	0.51
LOI	3.3	4.3	3.52	3.62	4.24	1.6	1.59	1.98	2.01	1.73	1.81
Total	99.07	99.11	99.1	99.32	99.64	99.31	98.07	98.93	99.44	99.31	99.34
Mg#	64	62	60	63	63	63	63	65	64	66	65
Li	21.7	28.3	21.5	24.4	30.3	21.8	20.7	25.3	24	13	12.4
Be	1.44	1.37	1.42	1.4	1.39	1.08	1.25	1.49	1.53	1.28	1.23
Sc	14.4	13.9	14.4	17.6	17.2	18.4	19.3	20.6	19.1	23.5	21
V	115	112	113	119	117	170	160	156	155	173	171
Cr	258	268	247	225	210	232	292	220	210	269	267
Co	13.8	22.3	20.3	16.5	18.2	32.9	21.7	27.6	26.9	35.5	35
Ni	114	118	116	128	114	138	117	136	127	149	148
Cu	7.5	7.6	7.8	23.7	24.6	18.9	41	79.9	66.3	32	36.2

(continued)

Table 3.2 (continued)

Sample	WGZ-4-1	WGZ-4-2	WGZ-4-3	WGZ-5-1	WGZ-5-2	WGZ-6-1	WGZ-6-2	WGZ-7-1	WGZ-7-2	WGZ-8-1	WGZ-8-2
Rock type	Lamprophyre dike			Lamprophyre dike		Dolerite dike		Lamprophyre dike		Dolerite dike	
	Wanggezhuang mafic dykes (Data are from Liu et al. 2020)										
Zn	76.9	72.5	76.4	70.9	52.7	80.2	74.6	97.2	92.3	81.1	81
Ga	19.2	19.1	18.9	18.2	17.2	18.8	19	18.7	18.8	18.2	17.9
Rb	76.9	71.5	75.5	72.5	57.4	71.1	62.4	78	76.5	65.1	62.5
Sr	680	582	674	586	543	498	528	667	670	670	650
Y	16.8	16.2	16.8	16.2	18	18	19.4	22.2	22.3	20.6	20
Zr	188	184	187	164	161	153	166	227	228	183	175
Nb	9.9	9.7	9.5	8.3	8.3	7	7.7	14.2	14.3	8.4	8.2
Mo	1.98	1.45	1.64	0.41	0.42	0.76	0.55	0.6	0.85	1.21	1.04
Cd	0.28	0.23	0.26	0.29	0.24	0.23	0.23	0.31	0.3	0.24	0.23
In	0.05	0.05	0.05	0.06	0.05	0.08	0.07	0.07	0.06	0.05	0.05
Cs	0.96	1.53	0.9	1.43	1.72	1.95	1.83	1.67	1.41	1.4	1.41
Ba	1817	1591	1837	1233	1487	1421	1444	1503	1534	1240	1218
La	72.9	70	73.5	73.4	62.4	49.7	55.7	92.4	87.4	75.2	73.3
Ce	136	129	137	147	124	95	105	162	156	147	144
Pr	14.5	13.7	14.5	16.3	14	10.4	11.3	18.5	18.1	16.1	15.8
Nd	53	50	53.1	60.7	53.8	39.8	42	67.5	67.1	61.2	59.8
Sm	8.12	7.72	8.04	8.97	8.44	6.56	6.92	10.26	10.34	9.69	9.58
Eu	2.59	2.44	2.56	2.54	2.59	1.98	2.11	3.1	3.15	2.86	2.84
Gd	7.51	7.29	7.58	7.73	7.73	6.48	6.62	9.63	9.54	8.88	8.77

(continued)

Table 3.2 (continued)

Sample	WGZ-4-1	WGZ-4-2	WGZ-4-3	WGZ-5-1	WGZ-5-2	WGZ-6-1	WGZ-6-2	WGZ-7-1	WGZ-7-2	WGZ-8-1	WGZ-8-2
Rock type	Lamprophyre dike			Lamprophyre dike		Dolerite dike		Lamprophyre dike		Dolerite dike	
	Wanggezhuang mafic dykes (Data are from Liu et al. 2020)										
Tb	0.76	0.73	0.76	0.79	0.82	0.74	0.77	0.98	1	0.93	0.91
Dy	3.64	3.5	3.64	3.68	4.02	3.94	4.03	4.86	5	4.61	4.55
Ho	0.65	0.65	0.66	0.65	0.73	0.75	0.79	0.9	0.92	0.84	0.83
Er	1.85	1.79	1.89	1.82	2.06	2.14	2.24	2.53	2.57	2.38	2.32
Tm	0.24	0.24	0.24	0.24	0.27	0.29	0.31	0.33	0.34	0.31	0.3
Yb	1.55	1.53	1.63	1.53	1.76	1.92	1.96	2.2	2.19	1.98	1.97
Lu	0.24	0.24	0.24	0.24	0.26	0.28	0.3	0.32	0.33	0.29	0.3
Hf	4.59	4.69	4.73	4.25	4.18	3.99	4.11	5.36	5.75	4.75	4.55
Ta	0.5	0.51	0.53	0.46	0.45	0.37	0.4	0.67	0.71	0.41	0.4
Pb	19.7	13.1	33.8	20.6	13.3	15.3	7.9	24.3	26.3	12.8	12.1
Bi	0.31	0.3	0.62	0.31	0.41	0.32	0.15	0.21	0.19	0.08	0.08
Th	10.2	9.3	9.9	8.6	7.9	7.1	8.3	9.6	9.9	9.8	9.2
U	2.21	2.34	2.56	2.06	1.98	1.51	1.69	1.89	1.93	2.06	1.97

Table 3.3 Whole-rock Sr–Nd–Hf isotopic compositions of the mafic dykes in the Jiaodong Peninsula

Sample	LYYJ-2-1	LYYJ-2-2	LYYJ-1-1	LYYJ-1-2	HYZW-3-1	HYZW-3-2	HYLGZ-1-1	LY01-1	LY01-2	LY01-3	LY02-1	LY02-3	LY02-4
	Haiyang mafic dykes (Data are from Liang et al. 2017)							Laiyang mafic dykes (Data are from Liang et al. 2018)					
$^{87}Rb/^{86}Sr$	0.53049	0.512248	0.135836	0.152492	0.495529	0.263181	0.13437	0.1561	0.1344		0.2192	0.1598	0.1932
$^{87}Sr/^{86}Sr$	0.709005	0.708952	0.707816	0.70777	0.708071	0.707676	0.708452	0.709113	0.709348		0.708842	0.708722	0.708639
2δ	0.000008	0.000008	0.000008	0.000008	0.00001	0.000008	0.000008	0.000017	0.000016		0.000018	0.000018	0.000019
$(^{87}Sr/^{86}Sr)_i$	0.7081	0.708079	0.707585	0.707509	0.707226	0.707227	0.708222	0.708847	0.709118		0.708468	0.708449	0.708309
$^{147}Sm/^{144}Nd$	0.115006	0.108317	0.113282	0.11662	0.123574	0.118796	0.101842	0.0927	0.0891		0.0946	0.0952	0.0969
$^{143}Nd/^{144}Nd$	0.511897	0.511898	0.511877	0.511884	0.511885	0.5119	0.511932	0.51163	0.511634		0.511757	0.511775	0.511762
2δ	0.000004	0.000004	0.000004	0.000005	0.000004	0.000005	0.000004	0.00001	0.000009		0.000009	0.000009	0.000009
$\varepsilon_{Nd}(t)$	−13.21	−13.09	−13.58	−13.49	−13.57	−13.21	−12.32	−18.07	−17.95		−15.62	−5.29	−15.57

Sample	MP-1-1	MP-1-3	MP-1-5	LYGC-2-3	LYGC-2-4	LLW50-1	LLW50-19
	Muping mafic dykes (Data are from Deng et al. 2017)					Lilong mafic dykes (Data are from Liang et al. 2020a)	
$^{87}Rb/^{86}Sr$	0.10614	0.09055	0.1094	0.19739	0.20519	0.385	0.163
$^{87}Sr/^{86}Sr$	0.70987	0.70993	0.70985	0.70878	0.70874	0.70953	0.710425
2δ	0.00001	0.00001	0.00001	0.00001	0.00001	0.000004	0.000006
$(^{87}Sr/^{86}Sr)_i$	0.70969	0.70977	0.70966	0.70845	0.70839	0.708817	0.710147
$^{147}Sm/^{144}Nd$	0.10286	0.10238	0.10494	0.09545	0.09391	0.101499	0.099316
$^{143}Nd/^{144}Nd$	0.51178	0.51176	0.51176	0.51191	0.51177	0.51188	0.511794
2δ	0.00001	0.00001	0.00001	0.00001	0.00001	0.00001	0.000009
$\varepsilon_{Nd}(t)$	−15.3581	−15.64	−15.6544	−12.5863	−15.2873	−13.2	−15

(continued)

Table 3.3 (continued)

Sample No	WGZ-4-1	WGZ-4-2	WGZ-4-3	WGZ-5-1	WGZ-5-2	WGZ-6-1	WGZ-6-2	WGZ-7-1	WGZ-7-2	WGZ-8-1	WGZ-8-2
	Wanggezhuang mafic dykes (Data are from Liu et al. 2020)										
$^{87}Rb/^{86}Sr$	0.3266	0.3544	0.3235	0.3574	0.3051	0.4122	0.3413	0.3374	0.3297	0.2808	0.2776
$^{87}Sr/^{86}Sr$	0.709575	0.709600	0.709581	0.709581	0.709592	0.709788	0.709718	0.709453	0.709459	0.709145	0.709134
±2σ	15	15	15	15	15	15	15	15	15	15	15
$(^{87}Sr/^{86}Sr)_i$	0.709575	0.709600	0.709581	0.709581	0.709592	0.709788	0.709718	0.709453	0.709459	0.709145	0.709134
$^{147}Sm/^{144}Nd$	0.0928	0.0935	0.0916	0.0895	0.0950	0.0996	0.0997	0.0919	0.0932	0.0958	0.0969
$^{143}Nd/^{144}Nd$	0.511616	0.511621	0.511637	0.511687	0.511703	0.511709	0.511711	0.511695	0.511700	0.511728	0.511744
±2σ	11	10	9	9	10	11	10	10	9	9	10
$(^{143}Nd/^{144}Nd)_i$	0.511616	0.511621	0.511637	0.511687	0.511703	0.511709	0.511711	0.511695	0.511700	0.511728	0.511744
εNd(t)	−19.9	−19.8	−19.5	−18.6	−18.2	−18.1	−18.1	−18.4	−18.3	−17.8	−17.4
T_{DM1}	1927	1932	1883	1790	1852	1920	1918	1816	1828	1833	1829
Lu	0.24	0.24	0.24	0.24	0.26	0.28	0.30	0.32	0.33	0.29	0.30
Hf	4.59	4.69	4.73	4.25	4.18	3.99	4.11	5.36	5.75	4.75	4.55
$^{176}Lu/^{177}Hf$	0.001342	0.001328	0.001339	0.001449	0.001592	0.001848	0.001901	0.001528	0.001475	0.001595	0.001709
$^{176}Hf/^{177}Hf$	0.282149	0.282147	0.282148	0.282344	0.282343	0.282254	0.282257	0.282257	0.282259	0.282268	0.282269
±2σ	0.000003	0.000003	0.000003	0.000003	0.000002	0.000003	0.000003	0.000003	0.000003	0.000003	0.000002
$(^{176}Hf/^{177}Hf)_t$	0.282149	0.282147	0.282148	0.282344	0.282343	0.282254	0.282257	0.282257	0.282259	0.282268	0.282269
ε_{Hf} (t)	−22.03	−22.10	−22.07	−15.14	−15.17	−18.32	−18.21	−18.21	−18.14	−17.82	−17.79
T_{DM1}(Ma)	1568	1570	1569	1297	1304	1440	1438	1423	1419	1410	1413

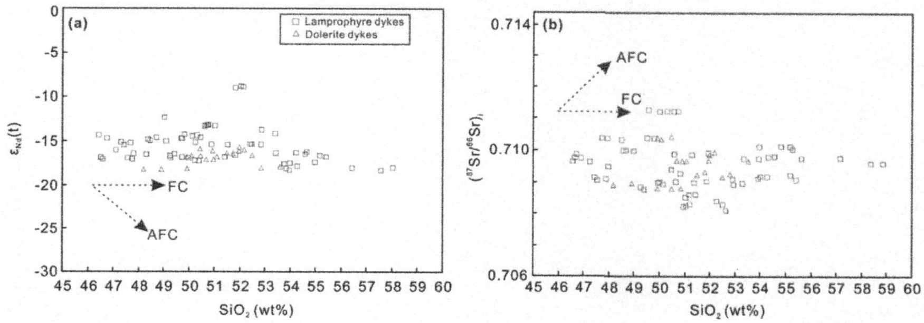

Fig. 3.7 Plots of $\varepsilon_{Nd}(t)$ versus SiO_2 **a** and $(^{87}Sr/^{86}Sr)i$ versus SiO_2 **b** for the Jiaodong mafic dykes. *Data sources* Ma et al. (2014, 2016), Cai et al. (2013, 2015), Liang et al. (2017)

rocks show some differences in the chemical composition, most of the samples still show similarly low SiO_2, high MgO, Cr, and Ni contents, and show high Mg# values. However, LILE enrichment, HFSE depletion, low $\varepsilon_{Nd}(t)$, and high $(^{87}Sr/^{86}Sr)_i$ ratios indicate that the Jiaodong mafic dykes may have the characteristics of crust-derived magma. These crust-derived characteristics may be related to assimilation and contamination of the crust, magma mixing, or to metamorphism in the source area (Yang et al. 2004; Tan et al. 2007). The trace element characteristics of the Jiaodong mafic dykes show that the content of Sr (>629 ppm), Rb (>32 ppm), and Ba (>817 ppm) is much higher than that of the North China Craton or the Yangtze Craton (Sr = 350–350 ppm; Rb = 59–79 ppm; Ba = 688–633 ppm; Gao et al. 1998). In addition, with the increase in SiO_2 content, the $(^{87}Sr/^{86}Sr)_i$ ratio does not change and the $\varepsilon_{Nd}(t)$ value does not gradually decrease (Fig. 3.7). These features rule out the possibility of significant crustal contamination (Ma et al. 2014). The granulite xenoliths contained in the Quaternary basalt found in the Nushan area of the southeastern North China Craton are considered to be typical representatives of the Mesozoic lower crust of that craton. The main features of the xenoliths are high Th/U, Zr/Hf, and Nb/Ta ratios, strongly negative whole rock $\varepsilon_{Nd}(t)$ anomaly, and very low $^{87}Sr/^{86}Sr$ value (Huang et al. 2012). However, these geochemical features have not been observed in the Jiaodong mafic dykes. Therefore, assimilation and contamination of crustal materials is not considered an important factor in the genesis of Jiaodong mafic dykes.

3.4.3 Fractional Crystallization

The elemental composition of the Jiaodong mafic dyke rocks showed similar characteristics, including a positive correlation between MgO–Cr and MgO–Ni (Fig. 3.8), which indicates that olivine and clinopyroxene may be the main components in magmatic evolution formed by separation of the crystalline phase. Although there is a negative correlation between MgO and Al_2O_3 (Fig. 3.8), there are no Sr and

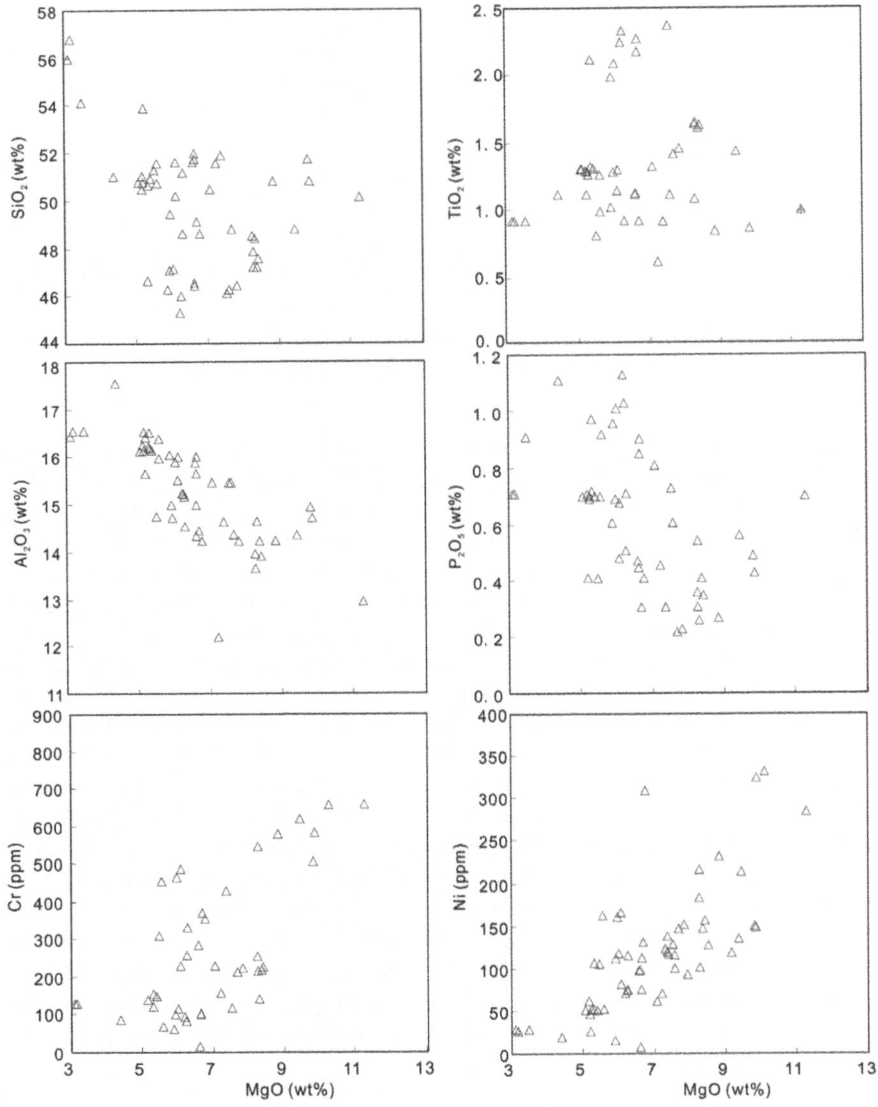

Fig. 3.8 MgO versus selected major and trace elements for the Haiyang mafic dykes. *Data sources* Ma et al. (2014, 2016); Cai et al. (2013, 2015); Liang et al. (2017)

Eu anomalies in the chondrite-normalised REE patterns and the standardised spider web maps of the original mantle, indicating that there is no obvious separation of plagioclase crystals (Fig. 3.5). No significant correlation between P_2O_5, TiO_2, and MgO was observed, which reflects the absence of obvious accessory minerals such as apatite and Fe–Ti oxides (Fig. 3.8). Therefore, the parent magma may have undergone the separation and crystallisation of olivine and clinopyroxene phases during its upward movement.

3.4.4 Magma Mixing: Evidence from Reversely Zoned Clinopyroxene Phenocrysts

3.4.4.1 Micro-Geochemistry of Clinopyroxenes

(1) **Major elements compositions of the clinopyroxenes**

Mineral observation under a microscope was performed at the Microscope Laboratory of the State Key Laboratory of Geological Process and Mineral Resources (Beijing), China University of Geosciences. The mineral electron probe microanalysis (EPMA) was performed at the electron probe laboratory of the State Key Laboratory of Geological Process and Mineral Resources (Beijing), China University of Geosciences. The electron probe microanalyzer instrument used is EMPA-1600, and the test conditions were as following: excitation voltage 15 kV, excitation current 10 A, beam spot diameter 1 μm, and standard samples were tested using 52 standard minerals from the US SPI company. The electron probe microanalyzer uses 60x magnification to acquire BSE images.

Group I clinopyroxene phenocrysts (IA and IB) have obvious core-mantle boundary zone structures (LY-2, LY-3, LY-3A, LY-3B; Fig. 3.9). As shown in Fig. 3.10, the composition of the cores of IA and IB clinopyroxenes are salite/augite and salite regions, respectively. Their rims have similar elemental compositions and can both be described as endiopside/diopside. Group II clinopyroxene is of enstatite/diopside composition. The core of the group I clinopyroxene from the porphyry has lower MgO and Al_2O_3, and higher Na_2O and FeO contents than the rims (Table 3.4; Fig. 3.9). In addition, according to the different element contents of the core, the group I clinopyroxene can be further divided into two subtypes. The core of the IA subtype (LY-3, LY-3A, and LY-3B) has and higher MgO and Al_2O_3 contents (LY-2) than core of the IB subtype, (Fig. 3.9). Group II clinopyroxene (MP-1, MP-2, LY-3C) has an acyclic band structure, and its element content is similar to the rims of the IA and IB groups, particularly in high Mg# value, high SiO_2, Al_2O_3, and Cr_2O_3 contents, as well as low contents of Na_2O and FeO (Table 3.4).

(2) **Trace elements compositions of the clinopyroxenes**

The in situ trace element content analyses of minerals were performed by LA-ICP-MS at the Mineral Deposit Genesis and Exploration Technology Research Center (OEDC), School of Resources and Environmental Engineering, Hefei University of Technology. The laser ablation system used was CetacAnalyte HE, and the ICP-MS was Agilent 7900. In the laser ablation process, helium was used as the carrier gas

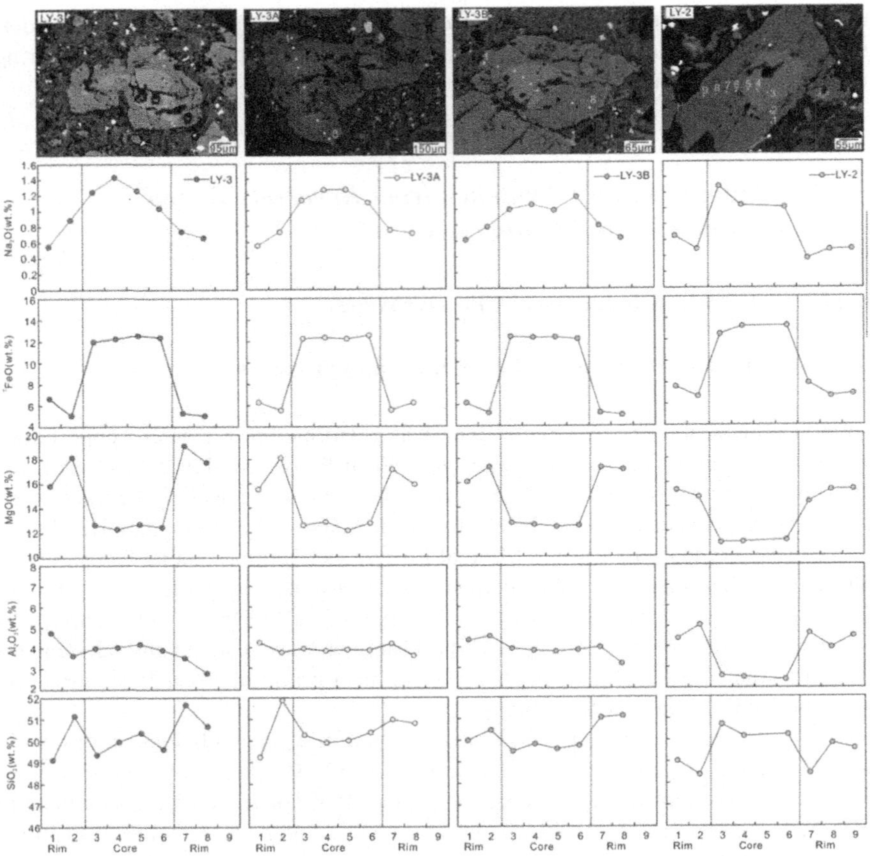

Fig. 3.9 Core-to-rim major element variations in the reversely-zoned clinopyroxene phenocrysts from the mafic dykes on Jiaodong Peninsula. Each column represents the composition of a single crystal as determined by EMPA

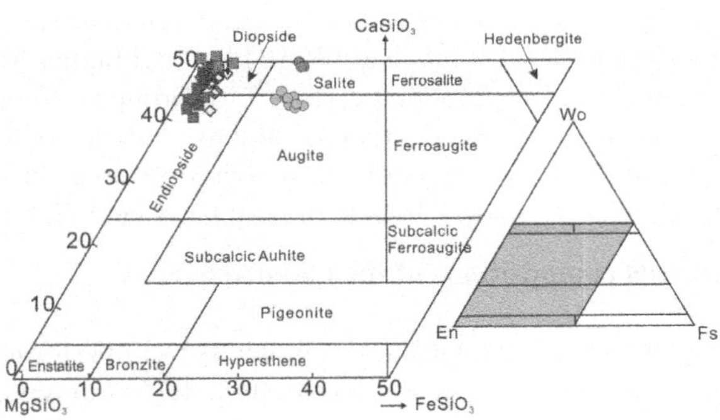

Fig. 3.10 $CaSiO_3$–$MgSiO_3$–$FeSiO_3$ diagram that shows the compositions of the clinopyroxene phenocrysts in the Jiaodong mafic dykes (pyroxene fields are after Morimoto et al. 1988)

Table 3.4 Major element concentrations (wt%) of clinopyroxene phenocrysts from the mafic dykes in the Jiaodong Peninsula (Data are from Liang et al. 2018)

Group	Group IA																								Group IB	
Sample	LY-3								LY-3A								LY-3B								LY-2	
Spot	Rim	Rim	Core	Core	Core	Core	Rim	Rim	Rim	Rim	Core	Core	Core	Core	Rim	Rim	Rim	Rim	Core	Core	Core	Core	Rim	Rim	Rim	Rim
SiO_2	49.1	51.2	49.4	50.0	50.4	49.6	51.8	50.7	49.3	51.9	50.3	49.9	50.0	50.4	51.0	50.8	50.1	50.5	49.5	49.8	49.6	49.7	51.0	51.1	49.0	48.4
TiO_2	0.89	0.44	0.76	0.69	0.62	0.78	0.36	0.50	0.85	0.38	0.63	0.66	0.61	0.55	0.29	0.49	0.49	0.48	0.52	0.62	0.62	0.59	0.24	0.46	0.68	1.22
Al_2O_3	4.74	3.63	3.98	4.03	4.18	3.89	3.51	2.79	4.22	3.73	3.92	3.80	3.86	3.83	4.16	3.56	4.32	4.57	3.95	3.85	3.80	3.88	4.01	3.20	4.40	5.05
^{T}FeO	6.59	4.94	12.0	12.3	12.6	12.4	5.18	4.92	6.26	5.50	12.3	12.3	12.2	12.5	5.50	6.18	6.17	5.22	12.3	12.2	12.3	12.1	5.23	5.01	7.62	6.70
MnO	0.19	0.08	0.38	0.23	0.03	0.19	0.07	0.01	0.09	0.11	0.17	0.47	0.31	0.21	0.25	0.17	0.30	0.38	0.05	0.21	0.26	0.21	0.16		0.25	0.42
MgO	15.2	17.3	12.3	12.0	12.4	12.1	18.2	16.9	15.5	18.1	12.6	12.8	12.1	12.7	17.2	15.9	16.0	17.3	12.7	12.5	12.4	12.5	17.2	17.0	15.4	14.8
CaO	22.2	19.6	19.1	18.3	18.9	18.6	19.6	20.9	22.0	19.0	19.2	18.6	18.9	19.2	19.5	21.6	21.7	19.6	18.6	18.4	18.7	18.9	19.9	20.4	22.1	22.4
Na_2O	0.54	0.88	1.24	1.43	1.26	1.03	0.73	0.65	0.55	0.72	1.12	1.25	1.25	1.09	0.74	0.70	0.61	0.77	0.99	1.05	0.98	1.15	0.79	0.63	0.65	0.49
K_2O		0.01		0.11	0.04	0.02		0.04		0.04	0.02	0.06				0.06					0.01				0.09	0.06
Cr_2O_3	0.63	0.97	0.11	0.12	0.18		0.79	1.12	0.52	0.83	0.01	0.04			0.43	0.67	0.54	0.63	0.22		0.24		0.56	1.23	0.29	0.25
Total	100	99.0	99.2	99.2	101	98.7	100	98.5	99.2	100	100	100	99.3	101	98.9	100	100	99.4	98.8	98.7	98.8	99.0	99.2	99.1	100	99.7
$Mg^{\#}$	80.4	86.2	64.7	63.5	63.6	63.6	86.2	86.0	81.5	85.4	64.6	64.9	63.9	64.4	84.8	82.1	82.2	85.5	64.7	64.6	64.2	64.7	85.4	85.8	78.2	79.7
Group	Group IB							Group II																		
Sample	LY-2							MP-2								MP-1									LY-3C	
Spot	Core	Core	Core	Rim	Rim	Rim	Rim	1	2	3	4	5	6	7	8	1	2	3	4	5	6	7	8	9	1	2
SiO_2	50.7	50.2	50.2	49.5	48.5	49.8	49.6	49.6	50.8	50.8	49.6	48.5	52.5	49.3	49.7	51.2	50.0	52.0	50.9	51.1	52.2	50.9	51.7	52.2	52.6	52.2
TiO_2	0.25	0.05	0.16	0.82	0.76	0.75	0.87	0.83	0.66	0.56	0.77	1.10	0.45	0.85	0.97	0.58	0.79	0.50	0.71	0.38		0.14	0.39	0.38	0.39	0.31
Al_2O_3	2.59	2.52	2.42	4.64	4.65	3.96	4.51	4.72	3.62	3.61	3.87	5.07	2.37	4.73	4.09	3.49	3.57	3.10	3.11	3.35	2.73	1.81	2.21	2.24	3.41	2.80
^{T}FeO	12.5	13.2	13.3	7.38	7.92	6.71	6.93	6.61	5.19	5.64	6.08	7.32	6.17	6.08	6.07	5.53	5.12	5.19	5.46	5.21	5.23	5.11	5.30	5.72	5.59	4.95

(continued)

Table 3.4 (continued)

Group	Group IB							Group II																		
Sample	LY-2							MP-2								MP-1									LY-3C	
Spot	Core	Core	Core	Rim	Rim	Rim	Rim	1	2	3	4	5	6	7	8	1	2	3	4	5	6	7	8	9	1	2
MnO	0.38	0.27	0.48	0.11	0.17	0.24		0.14		0.12	0.17		0.21	0.36	0.11	0.19	0.28	0.02	0.09	0.29		0.11	0.24	0.21	0.22	0.19
MgO	11.0	11.1	11.3	15.6	14.4	15.4	15.4	15.7	16.5	16.7	16.1	15.0	17.8	16.0	15.8	16.6	16.3	17.3	16.7	16.7	17.6	17.8	16.9	17.1	19.0	17.9
CaO	21.2	21.2	21.2	21.1	22.1	22.3	22.1	21.5	21.8	21.7	21.8	21.5	20.2	20.9	21.9	21.9	21.9	22.1	21.9	22.0	21.9	20.3	21.6	21.6	18.1	20.7
Na_2O	1.28	1.04	1.01	0.70	0.37	0.48	0.49	0.43	0.42	0.48	0.38	0.43	0.44	0.48	0.36	0.55	0.45	0.41	0.51	0.48	0.37	0.30	0.24	0.35	0.66	0.51
K_2O	0.11			0.13		0.05	0.05				0.02			0.02			0.00	0.02		0.02			0.07		0.02	
Cr_2O_3	0.05		0.09	0.23	0.26	0.37	0.38	0.74	1.16	0.87	0.51	0.12	0.51	0.48	0.14		1.04	0.94	0.89	0.95	0.98	0.38	0.31	0.22	0.49	0.67
Total	100	99.6	100	100	99.1	100	100	100	100	100	99.4	99.0	101	99.1	99.1	100	99.4	102	100	101	101	96.9	98.9	100	101	100
$Mg^{\#}$	61.1	59.9	60.2	79.0	76.4	80.3	79.9	80.9	85.0	84.1	82.6	78.5	83.7	82.4	82.3	84.2	85.0	85.6	84.5	85.1	85.7	86.1	85.0	84.2	85.8	86.6

and argon was used as the compensation gas to adjust the sensitivity. The two were mixed through a T-shaped joint before entering the ICP segment. Each time-resolved analysis dataset included a 40 s blank signal and 40 s sample signal. Offline data processing included selection of samples and blank signals, instrument operating instructions, sensitivity drift correction, and element content determination using ICPMSDataCal software (Liu et al. 2010). Detailed instrument operation conditions and data processing methods are the same as those reported by Siyuan et al. (2017) and Fangyue et al. (2017). The mineral trace element content was determined using multiple reference glasses (NIST610, NIST612, NIST614, GSC-1G, GSD-1G, and BCR-2G) as external standards, without using internal standards (Liu et al. 2010).

The trace element contents of clinopyroxene phenocrysts are shown in Table 3.5. The core of group IA clinopyroxene has low contents of Cr, Ni, and Sr (Fig. 3.11) and shows a wide range in $\sum$REE contents (51–142 ppm). Furthermore, it shows enrichment in LREEs ($(La/Yb)_N = 3.91-8.09$), weak Eu anomalies ($Eu/Eu^* = 0.56–0.67$; Fig. 3.11a), depletion in HFSEs (Nb, Ta, Ti), and a low Nb/Ta ratio (0.018–0.034 ppm; Fig. 3.12b). In contrast, the core of clinopyroxene in group IB has lower Cr and Ni contents (Fig. 3.11), as well as lower trace element and $\sum$REE (8 ppm) contents. The core of IB clinopyroxene shows depletion in LREEs [$(La/Yb)_N = 0.55$] and HFSEs (Nb, Ta, Ti; Fig. 4.18d), but no obvious Eu anomaly (Fig. 3.12c). On the other hand, the clinopyroxene phenocrysts of groups IA and IB have similar trace element compositions, such as low REE contents (30–39 ppm), enrichment in LREEs ($(La/Yb)_N = 2.29–3.63$), high Cr and Ni contents, and depletion in HFSEs (Nb, Ta, Zr, Hf, Ti) (Fig. 3.12a–d). Group II clinopyroxene phenocrysts also have high Cr, Sr, and Ni contents (Fig. 4.17), low REE contents (36.7–54.4 ppm), and show enrichment in LREEs ($(La/Yb)_N = 1.93–4.36$). No obvious Eu anomalies ($Eu/Eu^* = 0.88–1.05$; Fig. 3.12e) nor depletion in HFSEs (Nb, Ta, Zr, Hf, Ti) (Fig. 3.12f) were observed.

(3) **Sr isotope compositions of the clinopyroxenes**

The in situ Sr isotope composition of clinopyroxene was obtained by MC-ICP-MS at the State Key Laboratory of Geological Process and Mineral Resources, China University of Geosciences (Wuhan). The VOLM (volume-optional and low-memory) device was used to eliminate isotope mass fractionation related to position effects. The isotope data were obtained from mass ^{82}Kr to ^{88}Sr with seven Faraday receiving cups in low-resolution static mode. Helium gas was used as a carrier gas and mixed with Ar to enter the ICP through a T-connector. Each time-resolved analysis dataset included a blank background signal of approximately 30 s (mainly to eliminate the background interference of Kr in the argon and helium gas flow, as well as the small but continuous Rb and Sr background effects from the rectangular tube and cone) and subsequent sample signal of about 50 s. The interference of ^{87}Rb to ^{87}Sr was corrected by measuring the intensity of the ^{85}Rb signal. The sample $^{87}Rb/^{85}Rb$ was calculated based on the reference glass StHs6/80-G value. Finally, by assuming $^{88}Sr/^{86}Sr = 0.1194$ as in the internal standard (Xu et al. 2013a), the fractionation of $^{87}Sr/^{86}Sr$ was corrected. The internal standard (glass CPX05 composed of Cpx, Sr = 518 ppm,

Table 3.5 Representative trace element contents (ppm) of clinopyroxene phenocrysts from the mafic dykes on Jiaodong Peninsula (Data are from Liang et al. 2018)

Group	Group IA															
Sample	LY-3					LY-3A						LY-3B				
Spot	Rim	Core	Core	Rim	Rim	Rim	Rim	Core	Core	Rim	Rim	Rim	Rim	Core	Core	Rim
Si	230635	230775	233720	241900	229700	234631	230635	233299	235449	238207	237412	238534	238862	232832	231289	233813
Ca	121292	128888	127103	128895	126681	122943	127147	121399	125991	130625	136696	115972	137326	125167	124702	118092
Ti	1413	2691	2511	1727	1597	1066	2681	2448	2380	2403	1868	1578	1701	2510	2528	1409
V	136	466	434	111	128	106	384	444	410	129	131	139	130	425	416	143
Cr	5614	103	128	3083	6186	5044	509	87.3	91.4	1188	6639	4892	5621	118	144	5626
Co	28.4	57.2	53.3	30.1	27.8	32.6	45.6	55.3	56.9	31.7	30.0	33.7	29.1	55.6	54.7	32.0
Ni	245	129	123	167	183	281	110	124	125	122	181	317	183	123	121	211
Rb	0.76	0.26	0.22	0.32	0.17	0.16	0.37	0.21	0.17	0.34	3.17	0.22	0.60	0.22	0.22	0.22
Sr	94.6	13.6	15.4	103	111	80.2	23.6	13.9	13.3	116	164	112	161	14.8	14.7	93.1
Y	6.89	12.3	11.6	9.65	8.18	7.01	9.77	10.9	11.0	15.9	8.98	8.87	9.26	11.7	11.1	8.22
Zr	10.2	66.5	64.7	21.6	14.4	10.4	48.6	59.9	55.0	52.7	20.0	23.2	19.2	63.8	62.2	13.0
Nb	0.11	0.14	0.12	0.06	0.06	0.05	0.10	0.11	0.10	0.03	0.36	0.34	0.20	0.12	0.10	0.10
Cs	0.03	0.14	0.14	0.12	0.09	0.03	0.03	0.12	0.10	0.13	0.12	0.12	0.13	0.12	0.12	0.11
Ba	22.4	1.38	1.98	18.0	4.00	2.78	0.49	0.24	0.36	10.51	135	25.6	65.7	0.84	0.46	3.66
La	3.28	5.80	5.71	4.16	3.93	2.18	4.57	5.67	5.65	8.75	6.88	4.80	5.30	5.85	5.80	3.38
Ce	12.6	23.0	22.5	16.7	15.5	7.91	17.7	22.1	21.3	32.2	23.0	17.2	17.8	22.3	21.8	13.0
Pr	1.97	3.46	3.38	2.85	2.55	1.41	2.60	3.23	3.12	5.33	3.40	2.75	2.86	3.28	3.24	2.19
Nd	11.6	16.1	15.0	15.6	13.4	9.07	13.7	14.7	14.3	27.0	16.7	14.5	15.0	15.7	15.6	12.2
Sm	2.89	3.87	3.62	3.87	3.33	2.57	3.35	3.30	3.41	6.30	3.76	3.42	3.54	3.74	3.64	3.06

(continued)

Table 3.5 (continued)

Group	Group IA															
Sample	LY-3					LY-3A						LY-3B				
Spot	Rim	Core	Core	Rim	Rim	Rim	Rim	Core	Core	Rim	Rim	Rim	Rim	Core	Core	Rim
Eu	0.74	0.82	0.77	1.15	1.10	0.66	0.68	0.71	0.75	1.87	1.07	1.05	1.06	0.79	0.75	0.92
Gd	2.05	3.23	3.05	2.86	2.79	1.97	2.72	3.05	2.95	5.13	2.93	2.86	2.59	3.15	2.98	2.32
Tb	0.25	0.49	0.43	0.40	0.35	0.25	0.38	0.46	0.45	0.71	0.42	0.39	0.37	0.47	0.50	0.31
Dy	1.47	2.81	2.74	2.19	1.89	1.71	2.51	2.53	2.66	3.65	2.30	1.96	2.15	2.75	2.73	1.80
Ho	0.28	0.53	0.52	0.41	0.35	0.27	0.41	0.46	0.47	0.72	0.39	0.38	0.38	0.51	0.48	0.33
Er	0.79	1.32	1.25	1.03	0.86	0.83	1.11	1.12	1.29	1.74	0.94	0.89	1.00	1.29	1.22	0.90
Tm	0.10	0.18	0.15	0.13	0.11	0.08	0.12	0.12	0.15	0.22	0.13	0.13	0.13	0.16	0.14	0.11
Yb	0.65	1.12	1.04	0.85	0.72	0.68	0.83	0.81	0.96	1.56	0.84	0.85	0.86	0.88	1.05	0.70
Lu	0.09	0.15	0.15	0.13	0.12	0.10	0.11	0.14	0.13	0.23	0.12	0.10	0.11	0.12	0.21	0.09
Hf	0.47	2.57	2.45	1.20	0.78	0.50	1.88	2.00	1.96	2.58	0.89	1.14	0.89	2.37	2.35	0.61
Ta	0.01	0.04	0.03	0.02	0.03	0.01	0.02	0.03	0.02	0.03	0.03	0.03	0.03	0.03	0.02	0.03
Pb	0.42	0.39	0.34	0.23	0.14	0.19	0.28	0.33	0.34	1.06	0.19	0.38	0.30	0.37	0.39	0.19
Th	0.14	0.07	0.10	0.06	0.04	0.13	0.04	0.06	0.05	0.23	0.40	0.26	0.33	0.06	0.06	0.09
U	0.02	0.03	0.03	0.03	0.02	0.02	0.02	0.03	0.03	0.03	0.06	0.03	0.03	0.03	0.03	0.02
La_N/Yb_N	3.63	3.71	3.93	3.49	3.92	2.29	3.96	5.00	4.22	4.02	5.89	4.06	4.43	4.79	3.96	3.45
Gd_N/Yb_N	2.62	2.38	2.42	2.77	3.21	2.39	2.72	3.10	2.54	2.72	2.89	2.79	2.49	2.97	2.35	2.73
Eu/Eu*	0.88	0.69	0.69	1.01	1.07	0.86	0.67	0.67	0.71	0.97	0.95	1.00	1.02	0.69	0.67	1.01

(continued)

Table 3.5 (continued)

Group	Group IA														
Sample	LY-3					LY-3A						LY-3B			
Spot	Rim	Core	Core	Rim	Rim	Rim	Rim	Core	Core	Rim	Rim	Rim	Rim	Core	Core	Rim
Sr/Y	13.7	1.11	1.33	10.6	13.6	11.4	2.41	1.28	1.21	7.26	18.3	12.7	17.4	1.27	1.33	11.3
La/Yb	5.06	5.17	5.48	4.87	5.47	3.19	5.53	6.97	5.89	5.61	8.21	5.66	6.17	6.68	5.52	4.81

Group	Group IB						Group II							
Sample	LY-2						MP-1			MP-2			LY-3C	
Spot	Rim	Rim	Core	Core	Rim	Rim	1	2	3	1	2	3	1	2
Si	226474	235823	232458	236992	234421	231336	239479	239479	243022	233009	233009	237412	230635	230635
Ca	147350	132278	142391	138134	139056	131414	147764	150167	155509	142742	144919	135667	117279	154152
Ti	4623	2041	474	530	2818	3226	2919	3453	3885	2593	3624	1579	1558	1690
V	198	104	169	182	192	161	216	257	273	182	233	144	156	124
Cr	1785	1985	67.4	51.2	715	2140	4861	4141	2458	6452	3259	4333	5930	3318
Co	32.8	25.9	38.4	48.2	36.6	31.1	32.0	32.7	36.7	30.6	31.5	35.7	30.2	25.7
Ni	129	114	37.6	45.0	90.0	118	156	139	133	173	132	185	186	144
Rb	0.23	0.89	0.64	0.35	0.78	0.17	0.13	0.14	0.29	0.15	0.09	0.19	0.18	1.05
Sr	168	99.3	22.9	35.8	139	136	62.0	68.5	67.6	62.6	58.5	56.7	92.0	124
Y	17.8	10.4	6.00	5.62	13.9	14.2	11.4	14.3	15.1	9.59	11.9	8.08	8.02	8.97
Zr	69.5	34.3	12.8	12.0	39.1	40.8	21.9	27.6	42.5	18.6	29.9	11.6	11.9	16.9

(continued)

Table 3.5 (continued)

Group	Group IB						Group II							
Sample	LY-2						MP-1			MP-2			LY-3C	
Spot	Rim	Rim	Core	Core	Rim	Rim	1	2	3	1	2	3	1	2
Nb	0.12	0.08	0.04	0.04	0.20	0.06	0.05	0.05	0.07	0.05	0.05	0.04	0.05	0.08
Cs	0.15	0.03	0.03	0.12	0.14	0.11	0.02	0.02	0.13	0.06	0.04	0.11	0.04	0.07
Ba	9.57	7.74	5.24	1.31	62.7	0.45	0.42	0.41	0.51	0.19	0.38	0.41	16.2	17.1
La	11.1	5.07	0.49	0.92	7.05	7.96	3.14	3.89	4.38	2.79	3.33	2.10	3.16	4.79
Ce	41.3	16.3	1.27	2.79	26.8	31.0	12.1	14.9	17.3	10.0	12.7	8.66	12.0	15.4
Pr	6.87	2.81	0.20	0.39	4.31	5.21	2.03	2.50	2.88	1.75	2.20	1.37	2.01	2.66
Nd	36.1	17.6	1.33	2.10	22.6	26.8	12.3	15.7	15.6	10.7	12.9	7.51	12.2	15.9
Sm	8.4	4.35	0.50	0.58	5.61	6.20	3.31	4.21	4.11	2.89	3.54	2.27	3.08	3.79
Eu	2.40	1.20	0.18	0.26	1.61	1.94	0.93	1.19	1.33	0.81	0.99	0.72	0.80	1.09
Gd	5.97	3.40	0.77	0.84	4.27	4.74	2.93	3.69	3.74	2.50	3.17	2.12	2.32	2.23
Tb	0.81	0.41	0.16	0.17	0.61	0.64	0.41	0.53	0.57	0.34	0.41	0.32	0.32	0.36
Dy	4.12	2.45	1.21	1.21	3.36	3.38	2.60	3.38	3.62	2.26	2.76	1.76	2.03	2.16
Ho	0.74	0.40	0.25	0.24	0.59	0.61	0.46	0.58	0.68	0.39	0.48	0.35	0.29	0.34
Er	1.84	1.19	0.72	0.63	1.44	1.44	1.28	1.75	1.55	1.13	1.33	0.79	0.95	0.98
Tm	0.23	0.13	0.14	0.09	0.19	0.19	0.16	0.20	0.20	0.13	0.16	0.12	0.11	0.10
Yb	1.46	1.01	0.64	0.53	1.38	1.25	1.06	1.45	1.30	0.93	1.17	0.72	0.81	0.79
Lu	0.19	0.13	0.09	0.08	0.17	0.17	0.14	0.16	0.17	0.11	0.15	0.11	0.09	0.09

(continued)

Table 3.5 (continued)

Group	Group IB						Group II							
Sample	LY-2						MP-1			MP-2			LY-3C	
Spot	Rim	Rim	Core	Core	Rim	Rim	1	2	3	1	2	3	1	2
Hf	3.20	1.58	0.71	0.51	1.57	1.84	1.11	1.46	1.78	0.99	1.58	0.51	0.68	0.82
Ta	0.04	0.01	0.01	0.03	0.03	0.03	0.01	0.01	0.03	0.01	0.01	0.02	0.01	0.01
Pb	0.26	0.45	0.91	1.01	0.47	0.12	0.16	0.09	0.10	0.32	0.22	0.07	0.22	0.21
Th	0.20	0.38	0.10	0.10	0.23	0.06	0.06	0.05	0.06	0.04	0.05	0.05	0.13	0.17
U	0.04	0.09	0.03	0.03	0.03	0.03	0.01	0.01	0.03	0.01	0.01	0.03	0.01	0.04
La_N/Yb_N	5.44	3.61	0.55	1.25	3.66	4.57	2.13	1.93	2.42	2.15	2.05	2.09	2.81	4.36
Gd_N/Yb_N	3.38	2.79	1.00	1.32	2.56	3.14	2.30	2.11	2.38	2.22	2.25	2.44	2.38	2.34
Eu/Eu*	0.99	0.92	0.90	1.15	0.97	1.05	0.90	0.90	1.02	0.90	0.88	0.99	0.88	1.05
Sr/Y	9.42	9.57	3.82	6.37	9.97	9.61	5.42	4.79	4.48	6.53	4.93	7.02	11.5	13.8
La/Yb	7.59	5.03	0.76	1.74	5.11	6.37	2.98	2.69	3.37	3.00	2.86	2.92	3.91	6.08

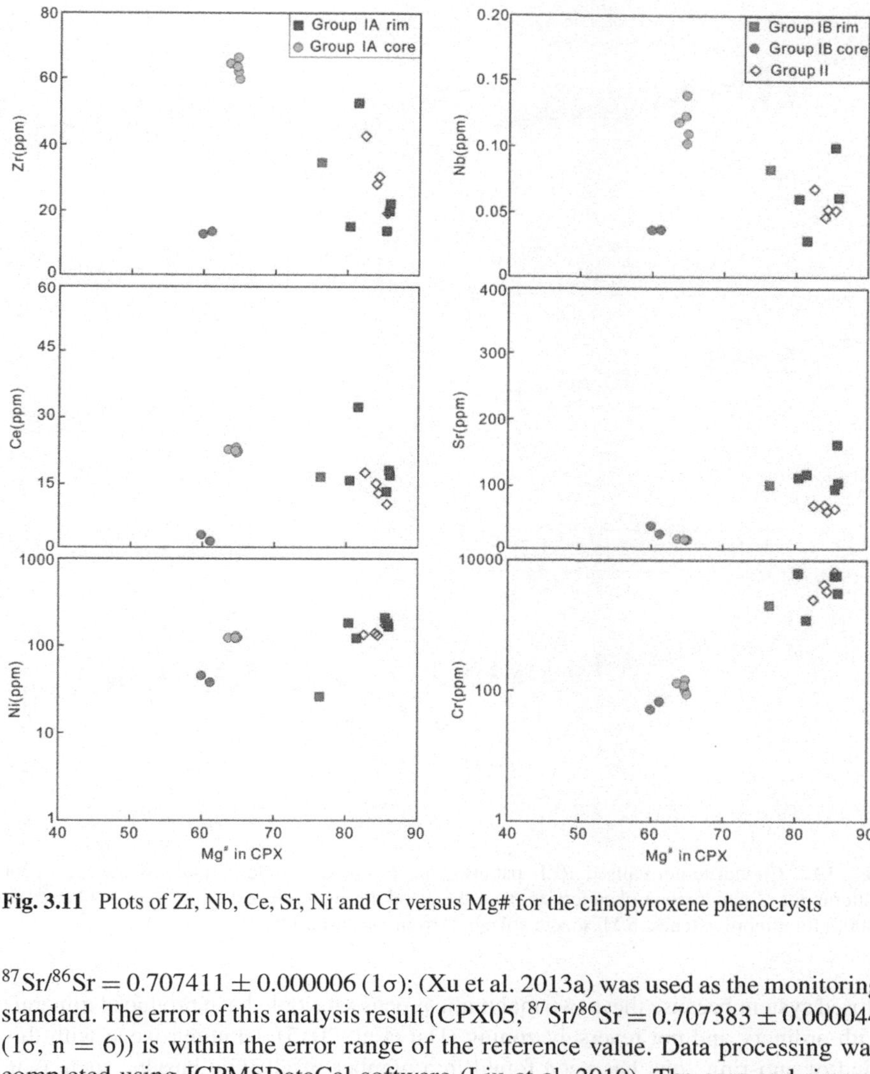

Fig. 3.11 Plots of Zr, Nb, Ce, Sr, Ni and Cr versus Mg# for the clinopyroxene phenocrysts

$^{87}Sr/^{86}Sr = 0.707411 \pm 0.000006$ (1σ); (Xu et al. 2013a) was used as the monitoring standard. The error of this analysis result (CPX05, $^{87}Sr/^{86}Sr = 0.707383 \pm 0.000044$ (1σ, n = 6)) is within the error range of the reference value. Data processing was completed using ICPMSDataCal software (Liu et al. 2010). The core and rims of group I clinopyroxenes (IA and IB core and rim) and group II clinopyroxene have similar in situ $^{87}Sr/^{86}Sr$ ratios (0.707647–0.708844) (Table 3.6; Fig. 3.13).

According to the geochemical characteristics of the whole rock, clinopyroxene and olivine are the main separated crystalline minerals in the Jiaodong mafic dyke rocks. The group I clinopyroxene phenocrysts in the mafic dykes have an obvious reversely structure. The core has a low Mg# value, high Na_2O and FeO contents, but the rim has higher Mg# value, higher Cr and Ni contents, and lower Na_2O and FeO contents (Fig. 3.9). This complex zoning structure provides evidence against contamination of magma by the earth's crust in an open system or separation and crystallisation in a

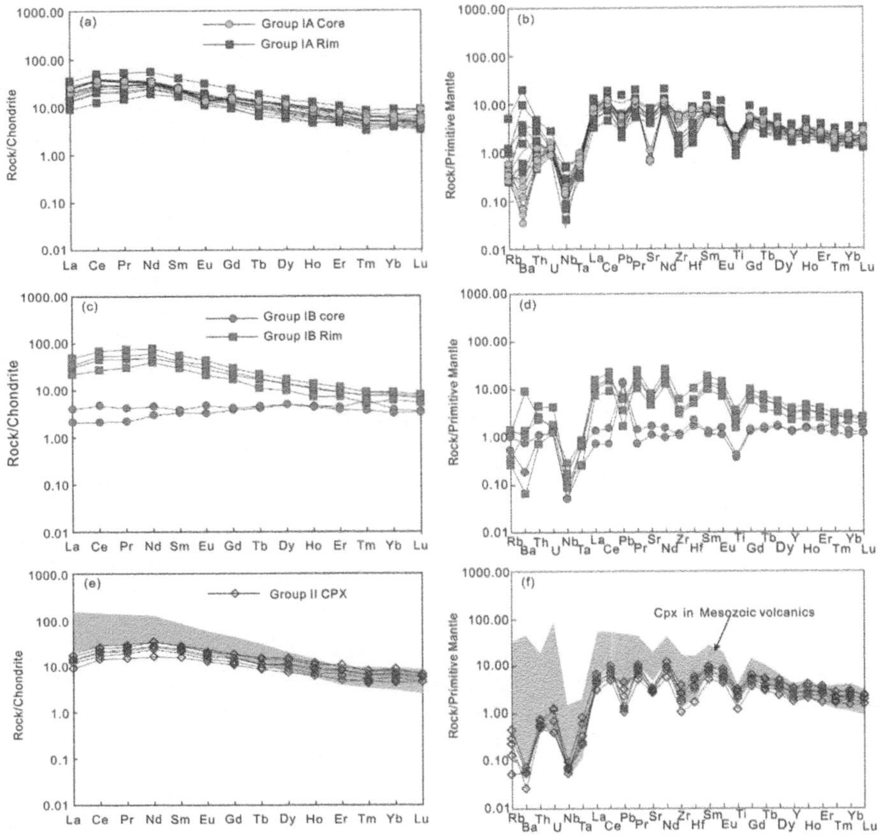

Fig. 3.12 Chondrite-normalised REE patterns and primitive-mantle-normalised trace element patterns for clinopyroxene phenocrysts from the Jiaodong mafic dykes. *Data sources* Gao et al. (2008) for clinopyroxenes in Mesozoic volcanics from the eastern NCC

closed system because these two magmatic processes would have produced minerals with ordinary and not reversely zoning (Dai et al. 2015). Clinopyroxene with this kind of anti-ring zone has been found in a number of alkaline basalts, such as in Uganda (Lloyd 1981), central France (Pilet et al. 2002), western Yunnan (Xu et al. 2003), and the eastern part of the North China Craton (Zhang et al. 2007; Gao et al. 2008). Two main mechanisms for the formation of this kind of anti-ring structures have been previously proposed: melt–mantle mixing reaction and magma mixing (Guo et al. 2007; Gao et al. 2008; Streck 2008; Muravyeva et al. 2014; Dai et al. 2015).

Table 3.6 Representative Sr isotopic compositions of clinopyroxene phenocrysts from the mafic dykes in the Jiaodong Peninsula (Data are from Liang et al. 2018)

Sample	Position	Rb	Sr	^{84}Sr	^{86}Sr	^{87}Sr	^{88}Sr	^{85}Rb	^{87}Sr/^{86}Sr	1σ	^{84}Sr/^{86}Sr	1σ	^{85}Rb/^{86}Sr	1σ
Group IA														
LY-3	Rim	63.0	1479	0.03245	0.59341	0.63082	5.14362	0.50634	0.708548	0.00004	0.056561	0.00007	0.901605	0.02633
	Core	1.02	185	0.00447	0.07427	0.05688	0.64426	0.00819	0.708669	0.00013	0.061706	0.00045	0.126236	0.00959
	Core	20.3	655	0.01455	0.26281	0.25476	2.27864	0.16281	0.708359	0.00005	0.057250	0.00020	0.644960	0.01925
	Rim	3.20	123	0.00339	0.04934	0.04582	0.42712	0.02572	0.707647	0.00020	0.071969	0.00102	0.496889	0.04720
LY-3A	Rim	13.5	467	0.01066	0.18735	0.17854	1.62366	0.10859	0.708249	0.00009	0.059185	0.00050	0.582673	0.02076
	Core	12.1	158	0.00381	0.06325	0.08459	0.54877	0.09705	0.708843	0.00032	0.061952	0.00061	1.448832	0.06602
LY-3B	Rim	25.2	972	0.02134	0.39005	0.36248	3.38249	0.20240	0.708537	0.00004	0.056497	0.00009	0.551152	0.01670
	Core	26.9	753	0.01632	0.30199	0.30455	2.61843	0.21643	0.708431	0.00007	0.055921	0.00018	0.742027	0.01858
	Core	1.34	77.3	0.00224	0.03102	0.02666	0.26857	0.01076	0.708535	0.00055	0.076151	0.00187	0.390660	0.02398
	Rim	34.3	814	0.01803	0.32644	0.34584	2.82950	0.27559	0.708508	0.00004	0.057004	0.00014	0.860686	0.01779
Group IB														
LY-2	Rim	50.3	1056	0.02295	0.42371	0.46756	3.67258	0.40407	0.708667	0.00004	0.056035	0.00006	0.979731	0.02018
	Core	1.41	270	0.00652	0.10817	0.08253	0.93778	0.01135	0.708384	0.00011	0.062027	0.00039	0.102784	0.00500
	Core	6.43	570	0.01296	0.22883	0.18570	1.98407	0.05166	0.708499	0.00006	0.058327	0.00025	0.246710	0.01182
Group II														
MP-1	1	0.81	205	0.00536	0.08230	0.06197	0.71342	0.00648	0.708616	0.00014	0.069016	0.00069	0.063648	0.00465

(continued)

Table 3.6 (continued)

Sample	Position	Rb	Sr	^{84}Sr	^{86}Sr	^{87}Sr	^{88}Sr	^{85}Rb	$^{87}Sr/^{86}Sr$	1σ	$^{84}Sr/^{86}Sr$	1σ	$^{85}Rb/^{86}Sr$	1σ
	2	17.0	340	0.00685	0.13647	0.15350	1.18521	0.13676	0.708844	0.00008	0.051798	0.00024	1.028557	0.01174
MP-2	1	0.98	153	0.00433	0.06121	0.04726	0.52980	0.00791	0.708134	0.00019	0.073080	0.00075	0.137586	0.00750
	2	13.3	220	0.00517	0.08807	0.10631	0.76414	0.10650	0.708190	0.00022	0.061370	0.00061	1.310536	0.13708
Standards														
CPX-05		0.22683	739.178	0.01657	0.29652	0.21414	2.56978	0.00182	0.707377	0.00005	0.05785	0.00012	0.00619	0.00004
CPX-05		0.27942	828.616	0.01869	0.33240	0.24010	2.88030	0.00224	0.707355	0.00004	0.05821	0.00010	0.00683	0.00003
CPX-05		0.20385	722.335	0.01607	0.28976	0.20920	2.51148	0.00164	0.707358	0.00005	0.05737	0.00015	0.00571	0.00004

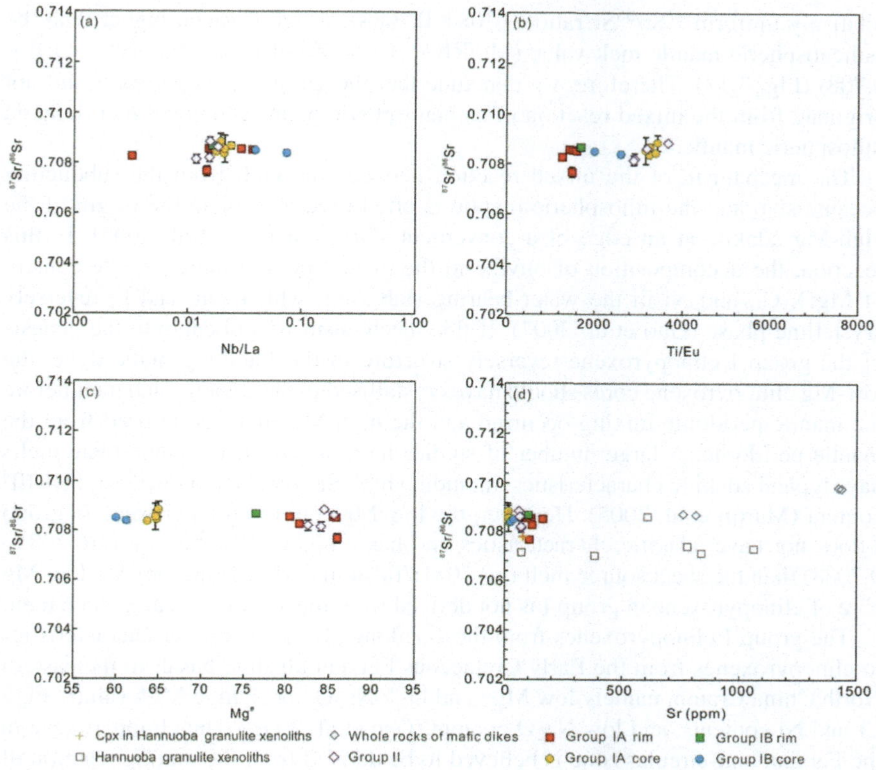

Fig. 3.13 Plots of $^{87}Sr/^{86}Sr$ ratios versus Sr, Nb/La, Ti/Eu and Mg# for the clinopyroxenes. *Data sources* $^{87}Sr/^{86}Sr$ ratios for the clinopyroxenes in the Cenozoic Hannuoba basalts are from Zhou et al. (2002), whole-rock data for the Cenozoic Hannuoba basalts are from Zhou et al. (2002), and whole-rock data for the Jiaodong mafic dykes are from Deng et al. (2017)

3.4.4.2 Melt–Mantle Interactions?

When the melt–mantle mixed reaction mechanism is used to explain the clinopyroxene reversely zoning, the melt often refers to the asthenospheric mantle, subducting oceanic slab, or delamination of continental crust (Pilet et al. 2002; Gao et al. 2008; Ronick and Renno 2010).

Group I clinopyroxene has similar geochemical characteristics (high Na core and low Na rim) as the green-core clinopyroxene in the alkaline basalts of central France (Pilet et al. 2002). These green-core clinopyroxenes are interpreted to be derived from the mixing reaction of asthenospheric mantle melt with metasomatic lithospheric mantle, and their core has a lower Nb/Th ratio than the rims (Pilet et al. 2002). The Jiaodong clinopyroxene is different from the above green-core clinopyroxene in that both the cores and the rims have negative Nb–Ta anomalies and low Nb content (Fig. 3.12b, d). Moreover, the Jiaodong clinopyroxene has a relatively uniform $^{87}Sr/^{86}Sr$ ratio (0.7084–0.7088), which is much higher than the asthenospheric mantle melt value (MORB < 0.704, Zindler and Hart 1986; OIB < 0.708) (Fig. 3.13). Therefore, we conclude that the group I clinopyroxene did not originate from the mixed reaction of asthenospheric mantle melt and metasomatic lithospheric mantle.

The mechanism of the mixed reaction between the melt from the subducting oceanic slab and the lithospheric mantle is often used to explain the origin of the high-Mg adakite at an edge of a convergent slab (Tsuchiya et al. 2005). In this reaction, the decomposition of olivine in the mantle peridot increases the content of MgO, Cr, and Ni in the water-bearing slab melt, which can lead to reversely crystalline phase (Guo et al. 2007). If this mechanism is applicable to the genesis of the group I clinopyroxene reversely structure in the Jiaodong mafic dyke, the low-Mg clinopyroxene cores should have crystallised in the oceanic slab melt before the mantle peridotite mixing occurred and the high-Mg edge crystallised from the mantle peridotite. A large number of studies have shown that oceanic plate melts have typical adakitic characteristics, including high Sr content and low Y and HREE content (Martin et al. 2005). However, the low-Mg core of clinopyroxene in group I does not have adakitic characteristics and has a higher $^{87}Sr/^{86}Sr$ ratio (0.7084–0.7088) than the sheet source melt (<0.7041; Tatsumi 2000). Therefore, the low-Mg core of clinopyroxene in group I is not derived from the adakitic oceanic plate melt.

The group I clinopyroxenes from the Jiaodong also have similar characteristics to clinopyroxenes from the Early Cretaceous Feixian alkaline basalt in the eastern North China Craton, namely low Mg# and high Na_2O cores, high Mg# values, high Cr and Ni contents, and low Na_2O content (Gao et al. 2008). The clinopyroxene in the Feixian anti-circular zone is believed to be derived from the mixing reaction of eclogite from the lower crust and the metasomatic mantle (Gao et al. 2008). However, compared with the Jiaodong Group I clinopyroxene ($Na_2O \leq 1.43$ wt%, $Al_2O_3 \leq 4.18$ wt%, Sr ≈ 18.7 ppm, and Sr/Y ≤ 2.4 wt%), Feixian reversely zone has a higher Na_2O ($\leq$2.4 wt%), Al_2O_3 ($\leq$6.9 wt%) and Sr (average 171 ppm) content, and a higher Sr/Y ratio (~30.1). Therefore, the reversely clinopyroxene of Jiaodong is not from the same source as the clinopyroxene of Feixian alkaline rock.

Based on these results, the clinopyroxene with the reversely zone of Jiaodong mafic dykes did not originate from the melt–mantle mixing reaction in the mantle source area. Moreover, the host mafic dyke has a relatively uniform $\varepsilon_{Nd}(t)$ (−15.6 to −12.6) and an initial $^{87}Sr/^{86}Sr$ ratio (0.7084–0.7098), indicating a uniform source.

3.4.4.3 Mixing Between Crust-Derived Felsic Melts and Mantle-Derived Basaltic Magma

(1) Source of clinopyroxene core of group IA

Group IA clinopyroxenes (LY-3, LY-3A, and LY-3B) have lower MgO and higher Na_2O and FeO^T contents in the bright core than in the dark rims (Fig. 3.9). According to the clinopyroxene–melt distribution coefficient $K_{D(Fe/Mg)}{}^{mineral/melt}$ =

0.275 ± 0.067 (Putirka et al. 2003), it is calculated that the core of clinopyroxene in group IA balances the Mg# of the original melt. The Mg# value is 37–39 (38 on average), which is much lower than that of peridotite (Mg# > 60; Gao et al. 2008) or pyroxenite/eclogite (Mg# = 85–65; Kogiso and Hirschmann 2001; Skjerlie and Patino Douce 2002), so the IA clinopyroxene core is not of mantle origin. Moreover, the Mg# value (Mg# = 64–65) of the clinopyroxene core of group IA is much lower than that of clinopyroxene in the basaltic magma (Mg# > 80, Gao et al. 2008), indicating that it is also not derived from the basaltic melt.

The in situ geochemical characteristics of the IA group clinopyroxene cores are similar to those of the clinopyroxene (FC-14) cores with the complex zonal origin of the early Cretaceous Fangcheng basalt in the southeastern NCC, which were interpreted to have been derived from newly accreted lower crust (Zhang et al. 2007). In this study, we compared the IA group clinopyroxene cores with the clinopyroxene in the Archean lower crustal granulite in Jiaodong (Liu et al. 2013) and clinopyroxene in the granulite xenoliths in the Meso-Cenozoic basalt in the eastern North China Craton (Chen et al. 2001; Liu et al. 2001; Zhou et al. 2002; Zheng et al. 2003; Huang et al. 2004; Ying et al. 2010). As shown in Fig. 3.14, the clinopyroxene cores of group IA and the granulite xenolithic clinopyroxene in the Cenozoic Hannuoba basalt of the eastern North China Craton have similar major and trace element characteristics and similar $^{87}Sr/^{88}Sr$ ratios (0.7084–0.7088; clinopyroxene $^{87}Sr/^{88}Sr \approx 0.7078$ in Hannuoba granulite xenoliths, host xenolith $^{87}Sr/^{88}Sr = 0.7070–0.7086$; Zhou et al. 2002) (Fig. 3.13). The granulite xenoliths in the Hannuoba basalt are mainly derived from products of basaltic underplating and subsequent fractional crystallisation at the base of the crust (Liu et al. 2001; Zheng et al. 2003). After multi-stage underplating, granulite-facies metamorphism occurred in the late Mesozoic (Zhang et al. 2007). Therefore, the core of clinopyroxene in group IA is derived from the newly-accreted lower crust formed by underplating of basaltic magma. The Sr isotope-enriched characteristics of the IA clinopyroxene cores may indicate that the underplating basaltic magma originates from an enriched mantle (Dai et al. 2016). In the Jiaodong Peninsula, arc-like mafic dyke rocks are enriched in Sr isotopes, LILEs and LREEs, and depleted in HFSEs and HREEs, indicating that they were formed by partial melting of the enriched lithospheric mantle (Ma et al. 2014; Deng et al. 2017). However, the IA clinopyroxene cores have lower REE and trace element contents than the rims and show negative Eu, Ba, and Sr anomalies (Fig. 3.12). Because elements such as Ba and Sr are mainly compatible with feldspar minerals, especially high-Na plagioclase (albite) (Keller et al. 2015), the negative Ba anomaly may indicate the separation of albite. In addition, the negative Eu and Sr anomalies indicate the simultaneous crystal of plagioclase.

(2) Origin of group IB clinopyroxene cores

The clinopyroxene phenocrysts of group IB also have the characteristics of reverse zoning, but their cores are obviously different from those of group IA clinopyroxene phenocrysts, with lower Mg#, Al_2O_3 (Fig. 3.9), ΣREE, and incompatible element contents (Fig. 3.12c–d). These characteristics indicate that the core of clinopyroxene porphyries in groups IB and IA have different source regions.

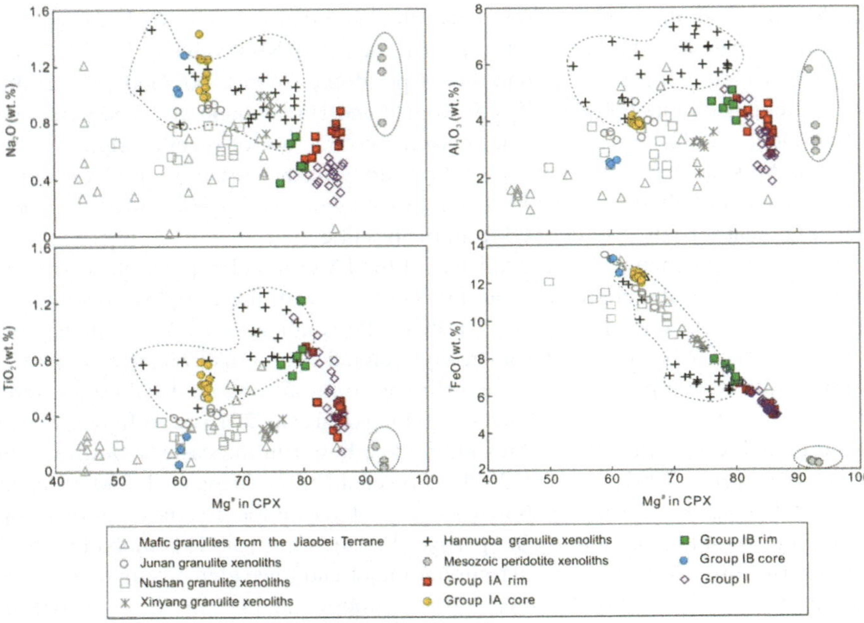

Fig. 3.14 Plots of Na_2O, Al_2O_3, TiO_2 and total FeO versus Mg# for the clinopyroxene (Cpx) phenocrysts *Data sources* are Liu et al. (2013) for clinopyroxenes in Archean lower-crust granulite from Jiaodong Peninsula, Ying et al. (2010) for clinopyroxenes in granulite xenoliths from Late Cretaceous basalts in Junan, Zheng et al. (2003) for Mesozoic volcaniclastic diatreme in Xinyang, Huang et al. (2004) for Cenozoic basalts in Nushan, Chen et al. (2001) for Cenozoic basalts in Hannuoba from the eastern NCC, and Zheng et al. (2005) for clinopyroxenes in peridotite xenoliths from the Cretaceous Xinyang basalts in the eastern NCC

The geochemical characteristics, such as low Mg#, $(La/Yb)_N$, Al_2O_3, and Sr contents of the clinopyroxene cores of the IB group are similar to those of the clinopyroxene Fe-rich cores of the high-Mg andesite in Yanji, Northeast China (Guo et al. al. 2007. Guo et al. (2007) suggested that the high-Mg core of clinopyroxene in the Yanji andesite originated from crust-derived andesitic-dacitic magma. According to calculations, the Mg# value of the balanced parent magma of the IB group clinopyroxene core is 34–35, so it is impossible that they have crystallised from mantle-derived or basaltic magma. In addition, the peraluminous magma formed in the mantle source region has $Al_2O_3 > 16$ wt% (Müntener et al. 2001; Streck et al. 2007), but the IB group clinopyroxene core has a low Al_2O_3 content (2.42–2.59 wt%), indicating the shallow crust as the possible source. Therefore, the low MgO, Al_2O_3, Cr, and Ni contents and high SiO_2 content of the group IB clinopyroxene cores indicate that they crystallised from a crust-derived acidic magma. In the Jiaodong Peninsula, the coeval crust-derived granites show initial $^{87}Sr/^{88}Sr$ ratios (0.708–0.710; Goss et al. 2010; Yang et al. 2012), which is similar to that of the IB group clinopyroxene cores.

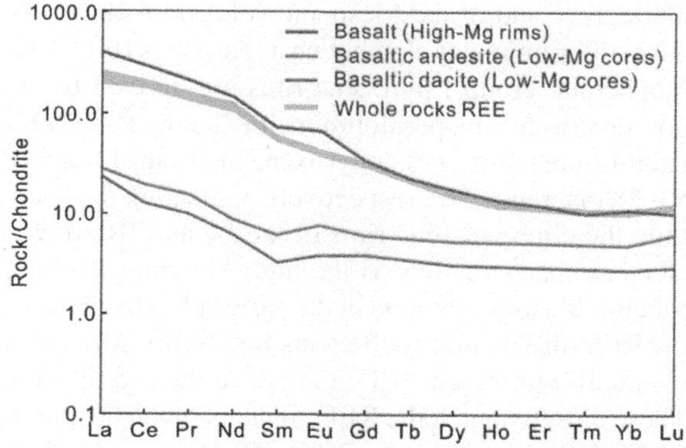

Fig. 3.15 Comparisons between the calculated REE contents of melts in equilibrium with the Group IB clinopyroxene cores and high-Mg clinopyroxene rims and the whole-rock REE concentrations of the Jiaodong mafic dykes

This is further evidence that the clinopyroxene cores were derived from the acidic magma of the crust source in the same period.

Based on the distribution coefficients of REEs in clinopyroxene in basalt, basaltic–andesite, and andesite–dacite rocks (Guo et al. 2007), we simulated and calculated the core composition of the group IB clinopyroxenes. The rims balance out the REE contents in the melt. As shown in Fig. 3.15, a mixture of balanced basaltic magma and andesitic-dacitic magma calculated from clinopyroxene can form Jiaodong mafic dykes. Therefore, it serves as an additional evidence that the core of the group IB clinopyroxene crystallised from crust-derived andesitic-dacitic magma.

(3) Origin of the Group I clinopyroxene rims

Group I (IA and IB) clinopyroxene rims have similar geochemical characteristics, including high Mg# values, high contents of Al_2O_3, Cr, and Ni, enrichment in LREEs and LILEs, depletion in HFSEs, and high initial $^{87}Sr/^{88}Sr$ ratios (Figs. 3.9, 3.10, 3.11 and 3.12). The ratios indicate that the rims of the IA and IB high-Mg clinopyroxenes crystallised from the same magma. Compared with the cores of IA and IB clinopyroxenes, the rims have higher Mg# values, Al_2O_3, Cr, and Ni contents, indicating that the core and rims originated from different sources. The high-Mg clinopyroxene rims indicate that an intrusion of a fresh and more primitive magma may have occurred in the process of magma evolution (Streck et al. 2007). Studies have shown that Na_2O in clinopyroxene is related to the crystallisation pressure (Adam and Green 1994; Gao et al. 2008), and that the distribution coefficient of Na between clinopyroxene and melt increases with an increase in the crystallisation pressure (Blundy et al. 1995; Putirka et al. 2003). The rims of group I clinopyroxenes have a lower content of Na_2O than the cores, indicating that it crystallises at a lower pressure and at a shallower depth. This is further evidence of a fresh and more primitive melt in the magma chamber, which envelops the early crystallised clinopyroxenes, that is, their low-Mg cores. The high Mg# (78–87) values and high contents of Al_2O_3 (3.96–4.05 wt%), Cr (1785–6639 ppm), and Ni (90–317 ppm) of the group I clinopyroxene rims indicate that they originate from a mantle-derived magma (Müntener et al. 2001; Streck et al. 2007). Based on the fact that the rims of the group I clinopyroxenes of the Jiaodong mafic dyke and of the Mesozoic volcanic rocks of the eastern North China Craton have different major element characteristics (Fig. 3.12), it can be concluded that the monoclinic group I pyroxene rims are derived from an evolved basaltic melt and not directly from a peridotite melt (Zheng et al. 2005). The Mg# values of the rims equilibrium melts of clinopyroxene in groups IA and IB are 58–68 (average ~65) and 53–58 (average ~56), respectively, indicating the source of basaltic magma. In addition, the clinopyroxene rims of the IA and IB groups have similar major and trace element characteristics as the high-Mg rims of clinopyroxenes in the Feixian and Sihetun Mesozoic basalts in the eastern North China Craton (Gao et al. 2008). Using the REE distribution coefficients for basaltic clinopyroxene, we inversely calculated the equilibrium melt REE contents in the rim of groups IA and IB clinopyroxenes. These closely match the REE contents in the host mafic dyke rock, which have high LREE/HREE ratios (Fig. 3.16). The highest Mg# of the rim equilibrium melt of clinopyroxene in groups IA and IB is also the same as that of the host mafic dyke. The $^{87}Sr/^{88}Sr$ ratios in the Jiaodong mafic dyke and the IA and IB group clinopyroxenes are also similar (Fig. 3.13). The outlined geochemical and mineralogical characteristics indicate that the rims of the group I clinopyroxenes have a direct genetic relationship with the host mafic dyke.

The $^{87}Sr/^{88}Sr$ ratios of the host Jiaodong mafic dyke, IA, and IB group clinopyroxene rims match closely. There are three possible mechanisms that can cause the homogenisation of the Sr isotopes: (1) crust-derived magma contamination (Streck 2008); (2) Sr isotope and trace element diffusion during magma mixing (Nakamura and Kushiro 1998; Perugini et al. 2008), and (3) different source regions that have similar $^{87}Sr/^{88}Sr$ ratios (Xu et al. 2013; Zou et al. 2014; Dai et al. 2015). Even if it has been proven that the magma mixing occurred during the evolution of the Jiaodong mafic dyke, the $^{87}Sr/^{88}Sr$ (0.7084–0.7097) of the clinopyroxene rims and the host rock reflect crust-derived magma contamination. The crust-derived material end-member

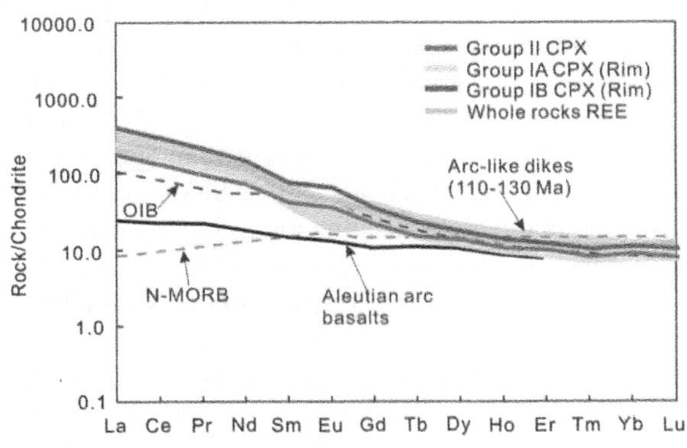

Fig. 3.16 Primitive-mantle-normalised trace element spidergrams showing the average melt compositions in equilibrium with high-Mg clinopyroxenes, as calculated by the clinopyroxene-melt partition coefficient $D_{Cpx/melt}$ (Adam and Green 2006). The primitive mantle trace element values are after McDonough and Sun (1995)

that could have contributed to the mixing, based on their Sr isotope composition, are the Early Cretaceous granite $((^{87}Sr/^{88}Sr)_i = 0.708–0.710$; Goss et al. 2010; Yang et al. 2012) or the freshly-formed lower crust ($^{87}Sr/^{88}Sr = 0.7070–0.7086$; Zhou et al. 2002). These are the only crust-derived materials that can be added at a 100% ratio to produce uniform $^{87}Sr/^{88}Sr$ ratios ($^{87}Sr/^{88}Sr \approx 0.7088$). However, if such formation mechanism had occurred, no reversely zone would have been produced. Therefore, the contamination of crust-derived materials has very little influence on the petrogenesis of the mafic dykes. Furthermore, if the similar Sr isotopes in the cores and rims of the groups IA and IB clinopyroxenes are due to element diffusion, the elements with higher diffusion rates (such as Rb and Li; Perugini et al. 2008) would have shown chemical equilibrium, which is not the case in the core and rims. Finally, the similar Sr isotope of clinopyroxene in the reversely zone of the Jiaodong mafic dyke may be derived from different sources with similar isotopes. Studies have shown that the early Cretaceous felsic intrusive rocks and lithospheric mantle-derived mafic dyke rocks in Jiaodong have similarly high Sr isotopic compositions (Chen et al. 2001; Zhang et al. 2002; Ma et al. 2014; Deng et al. 2017). The Jiaodong lithospheric mantle-derived mafic dykes originate from an enriched lithospheric mantle (Ma et al. 2014; Deng et al. 2017). The high Sr isotope in the core of IA and IB clinopyroxenes groups is sourced from continental crust. However, the high-Mg clinopyroxene rims and host mafic dyke have enriched Sr isotope compositions, which are likely to be related to the same enrichment.

In summary, the group I clinopyroxene showing reversely zoning indicates magma mixing, that is, the high-Mg magma (reflected in the rims of the crystalline high-Mg clinopyroxenes) mixed with crust-derived andesitic-dacitic magma sourced (IB core) or newly formed lower crustal material (IA core).

3.4.5 Nature of the Mantle Source of the Mafic Dykes

The Jiaodong mafic dyke is characterised by low SiO_2, high MgO content and Mg# value, enrichment in compatible elements, indicating that the vein rock is sourced from an ultramafic mantle. The geochemical characteristics and zircon Hf isotopic characteristics of the Jiaodong mafic dykes indicate that it may be derived from a single magma rather than a mixture of magmas. The Sr–Nd isotopic characteristics of the dyke rocks are different from those of oceanic island-like basic (OIB) dike rocks (~120 Ma) in the Jiaodong Peninsula, which is interpreted to be derived from the asthenospheric mantle (Ma et al. 2014). The Jiaodong mafic dyke is not similar to mid-ocean ridge basalt (MORB) or ocean island basalt (OIB) (Figs. 3.4 and 3.5). Early studies on Hf isotopes, for example, on zircons in southern Africa, Siberia, and Australian Kimberlite report lower radioactive Hf isotopes contents and $\varepsilon_{Hf}(t)$ values that reach -16. Griffin et al. (2000) considered that the lithospheric mantle of the craton subcontinent with a low Lu/Hf ratio (≤ 0.15), can lead to low $\varepsilon_{Hf}(t)$. The $\varepsilon_{Hf}(t)$ values of the Jiaodong mafic dyke range from -15.6 to -23.6, indicating a lithospheric mantle source. It is not only different from the depleted mantle source of

MORB but also from the lower crust of the North China Craton at the late Mesozoic in terms of low $\varepsilon_{Hf}(t)$ values (from −22.5 to +56.6; Yang et al. 2005). In addition, the negative $\varepsilon_{Hf}(t)$ value of the Jiaodong mafic dyke is similar to the $\varepsilon_{Hf}(t)$ (−26 ± 0.6) of the mantle-derived enclave in the Shijiusuo monzonitic granite (127 ± 2 Ma) in the Sulu orogen. The zircon $\varepsilon_{Hf}(t)$ value of the Jiaodong mafic dykes falls between 1.6 and 2.5 Ga on the crustal evolution line, which indicates that these dykes originated from ancient North China Craton lithospheric mantle (Liang et al. 2017). In addition, the Cretaceous Fangcheng basalt (~120 Ma) in the Luxi area shows high $(^{87}Sr/^{86}Sr)_i$ and low $\varepsilon_{Nd}(t)$ values, and it has been considered to be derived from the lithospheric mantle of North China Craton (Zhang et al. 2002). Due to the compositional similarities with the basalt, the Jiaodong mafic dyke may have also been derived from the lithospheric mantle of North China Craton.

Significant enrichment in LILEs, LREEs, and high K_2O content reflect that the formation and origin of the Jiaodong mafic dyke requires a mantle source enriched in these components. Hornblende and phlogopite are the most common volatile minerals that also represent the main reservoirs of LILEs in the lithospheric mantle (Ionov et al. 1997). In a melt equilibrium system, phlogopite end-members show higher Rb/Sr (>0.1) and low Ba/Rb (<20) ratios than the hornblende end-members (Furman and Graham 1999). The Jiaodong mafic dyke shows a wide range of Rb/Sr (0.04–0.18) and Ba/Rb (5–34) ratios, indicating that amphibole and phlogopite coexist in the melting source area (Fig. 3.17a). The K/Yb and Dy/Yb correlation diagrams can be used to evaluate constraints of the mantle source composition and to establish the degree of the partial melting (Fig. 3.17b; Duggen et al. 2005). Partial melting in the stable region of garnet generally produces higher Dy/Yb ratios (>2.5) than melting in the stable region of spinel (Duggen et al. 2005; Ma et al. 2014). The Jiaodong mafic dyke shows Dy/Yb ratios in the range 1.93–2.52 (average 2.16), falling between the partial melting curves of the garnet phase amphibole peridotite

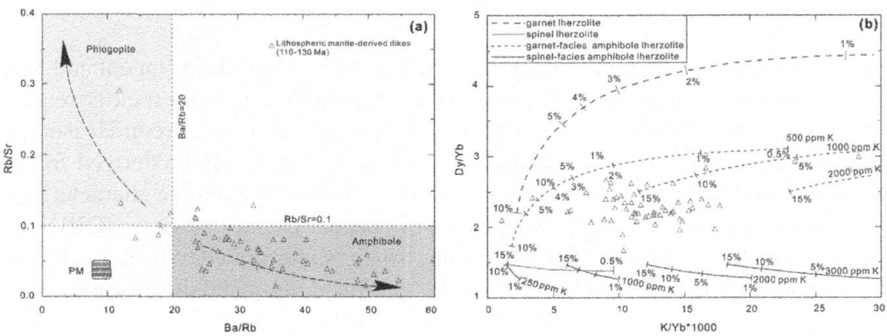

Fig. 3.17 **a** Rb/Sr versus Ba/Rb diagram (Furman and Graham 1999) and **b** plot of K/Yb versus Dy/Yb for the Haiyang mafic dykes. Melting curves for garnet lherzolite, spinel lherzolite, garnet-facies phlogopite lherzolite, garnet facies amphibole lherzolite, and spinel-facies amphibole lherzolite are from Duggen et al. (2005). *Data sources* Ma et al. (2014, 2016), Cai et al. (2013, 2015), Liang et al. (2017)

and the spinel-phase amphibole peridotite. The narrow Dy/Yb ratio range also fits with the lower HREE abundances, indicating that the source area garnet did not undergo a large amount of partial melting (Yang et al. 2007). The narrow range of K/Yb values reflects a low degree of partial melting. These characteristics indicate that the main source of the mantle of the Jiaodong mafic dyke is composed of amphibole- and phlogopite-bearing lherzolites in the spinel–garnet transition zone.

3.5 Mantle Source Were Enriched by the Subduction-Related Metasomatism

Based on the discussion and results outlined above, it can be seen that the source of the Jiaodong mafic dyke is likely to be enriched lithospheric mantle, such as enrichment of LILEs and LREEs, depletion of HFSEs, high $^{87}Sr/^{86}Sr$ ratios and low $\varepsilon_{Nd}(t)$ values (Figs. 3.4 and 3.5). The negative Nb, Ta, and Zr–Hf anomalies are probably related to the subduction experience in the source area (e.g., Thirlwall et al. 1994; Elliott et al. 1997; Ma et al. 2014; Cai et al. 2015). Moreover, the enriched lithospheric mantle that they are considered to originate from, was affected by metasomatism (Guo et al. 2004; Ma et al. 2014).

Zhang et al. (2002) proposed that the lithospheric mantle of the North China Craton shows a clear distinction in the composition of the rocks formed in the Mesozoic and Paleozoic. The characteristics of the mantle xenoliths in the Paleozoic kimberlites are high Mg# (>92) of olivine and high $^{87}Sr/^{86}Sr$ and $^{206}Pb/^{204}Pb$ ratios of the whole rock composition (Zheng and Lu 1999). This indicates that the ancient cratonic lithospheric mantle is largely covered by the Mesozoic lithosphere with high $^{87}Sr/^{86}Sr$ and low $^{143}Nd/^{144}Nd$ ratios and relatively low radiogenic Pb isotope values (Zhang et al. 2002). This opposes the formation of the enriched lithospheric mantle by multiple subduction-related events that occurred in the Late Archean to Proterozoic. Based on the study of Cretaceous dyke rocks in the Guocheng gold deposit in the Jiaodong Peninsula, Tan et al. (2013) reported that the Mesozoic mantle metasomatism did not originate from the subduction of the Yangtze plate because the Pb isotopic characteristics of the Yangtze plate under the subducted crust are of radiogenic origin. Liu et al. (2009) studied the mafic dyke of the gold deposit in the northern slope of the Jiaodong Peninsula, which is different from the lithospheric mantle of the Yangtze plate in terms of Pb isotopic characteristics, and excluded the participation of the lithosphere of the Yangtze plate. Mafic dyke rocks with these characteristics can also be found in the Linglong and Penglai areas of the Jiaodong Peninsula (Ma et al. 2014). Therefore, the enrichment of the lithospheric mantle cannot be attributed to the participation of the continental crust of the subducting Yangtze plate. In addition, the Nb/U ratio is not easily affected by partial melting or subsequent fractional crystallisation, and the Nb/U ratio range of magma often reflects its source area (Hofmann 1988; Xu et al. 2005; Ma et al. 2014). The Nb/U ratio of the Jiaodong mafic dyke rocks is 6.4–21.9, which is significantly lower than that of MORBs and OIBs (Nb/U

= 47 ± 7, Hofmann et al. 1986) and the lower crust (Nb/U ≈ 25; Rudnick and Gao 2003). These characteristics indicate that the enriched mantle source area is probably not related to the lower crust of the North China Craton nor does it support the aforementioned insignificant crust-mantle interaction.

Alternatively, the enriched mantle source of the Jiaodong mafic dykes was possibly related to metasomatism of the Paleo-Pacific plate during Mesozoic subduction. The ratios of some incompatible elements (such as Th/Yb, Ba/La, Hf/Sm, and Zr/Hf) can be largely affected by crustal contamination, and therefore used as an effective tracer to estimate the degree of mantle metasomatism (Ben et al. 1989; Ionov et al. 1993; Prelević et al. 2012). On the Th/Yb–Ba/La diagram (Fig. 3.18a), the Jiaodong mafic dyke shows a low and limited Th/Yb ratio range and an increase in the Ba/La ratio range. This indicates that the source area has imported fluids derived from the plate (Woodhead et al. 2001; Liu et al. 2014). In addition, Zr, Hf, and Sm all show similar geochemical behaviours during partial melting (McDonough 1990), and deep-sea sediments and ocean muddy rocks also exhibit low Hf/Sm ratio ranges (Liu et al. 2014). Therefore, the appearance of samples with low Hf/Sm but high Zr/Hf ratios may reflect that the mantle source area has undergone carbonate metasomatism (Fig. 3.18b; Ionov et al. 1993). The Jiaodong mafic dyke shows a fairly narrow range in Zr/Hf ratios and a large range in Hf/Sm ratios, which deviates from the Hf/Sm chondrite standardisation value of 0.69 (Fig. 3.18b). This difference in values is indicative of carbonate metasomatism in the mantle source area. In other words, the Jiaodong mafic dyke comes from the enriched lithospheric mantle altered by the carbonate metasomatism of the subducted slab. In addition, these dykes have higher Th contents and Th/La ratios than those of MORB (Fig. 3.18c) and do not exhibit adakitic features (Fig. 3.18d), which is inconsistent with the melts derived from oceanic slab material (Hermann and Rubatto 2009). However, the Jiaodong mafic dykes are characterized by high Rb/Y, at 2.24–3.35, and relatively low Nb/Y, at 0.36–0.53 (Fig. 3.18e–f). This further implies that the mantle source has been metasomatized through subduction-related fluids from an oceanic slab. Moreover, the whole-rock geochemical characteristics further support the marine sediment-derived materials in the mantle source.

3.5.1 Enriched Features of High-Mg Clinopyroxene Phenocrysts

The Group II clinopyroxene phenocrysts have similar geochemical features to the Group I clinopyroxene phenocrysts rims, including high values of $Mg^{\#}$, high contents of MgO, Al_2O_3, Cr, Ni and Sr, low contents of Na_2O, and similar values of $^{87}Sr/^{88}Sr$ (Figs. 3.9, 3.13 and 3.19). On the chondrite-normalised REE patterns and the primitive-mantle-normalised trace element spidergrams, the Group II clinopyroxenes and Group I clinopyroxene rims display similar patterns, with enrichments in LREEs and depletions in HREEs and HFSEs (Fig. 3.20). These observations suggest

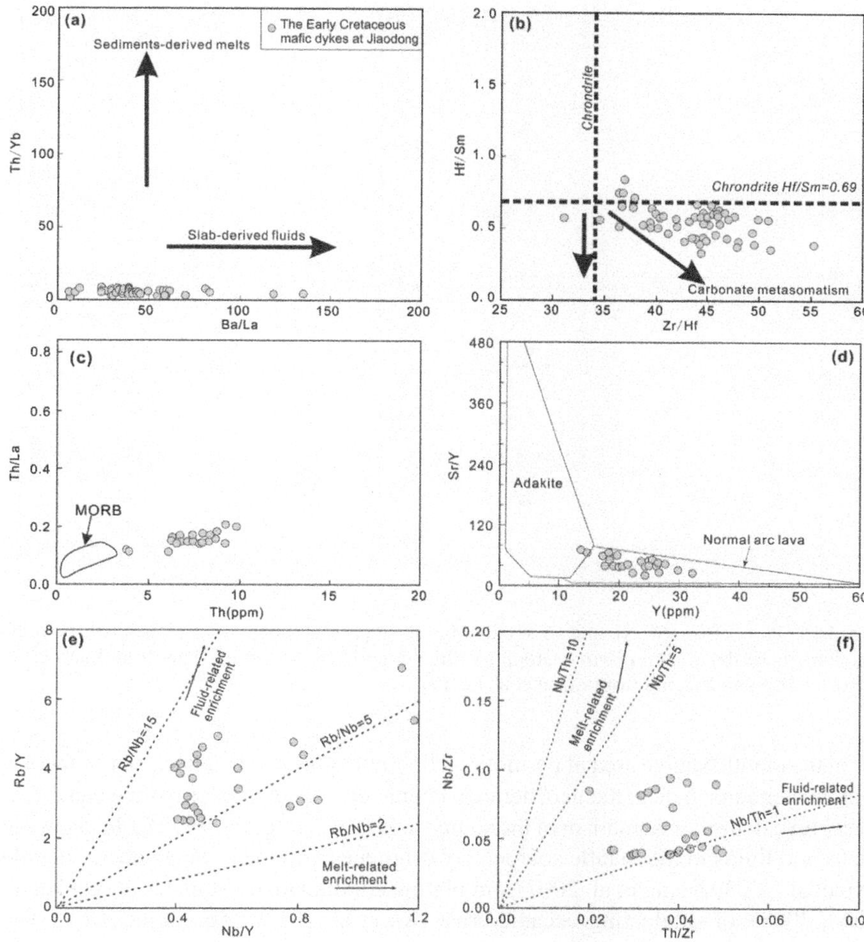

Fig. 3.18 Plots of **a** Th/Yb versus Ba/La; **b** Hf/Sm versus Zr/Hf; **c** Th versus Th/La (MORB data are from Niu and Batiza 1997); **d** Sr/Y versus Y (after Martin et al. 2005); **e** Rb/Y versus Nb/Y; **f** Nb/Zr versus Th/Zr for the Jiaodong mafic dykes. *Data source* The Early Cretaceous mafic dykes at Jiaodong are from Ma et al. (2014), Liang et al. (2017, 2019)

that the Group II clinopyroxenes and the Group I clinopyroxene rims (both high-Mg clinopyroxenes) were derived from the same source, namely lithospheric mantle. In our discussion above, we excluded magma mixing from having any influence on the enriched Sr isotopic compositions. We conclude, therefore, that the enriched Sr isotopic features probably represent metasomatism of the source area, and this conclusion is supported by the strong depletions in HFSEs in the high-Mg clinopyroxenes (Liu et al. 2005; Xu R. et al. 2013; Zou et al. 2014; Zeng et al. 2016) (Table 3.7).

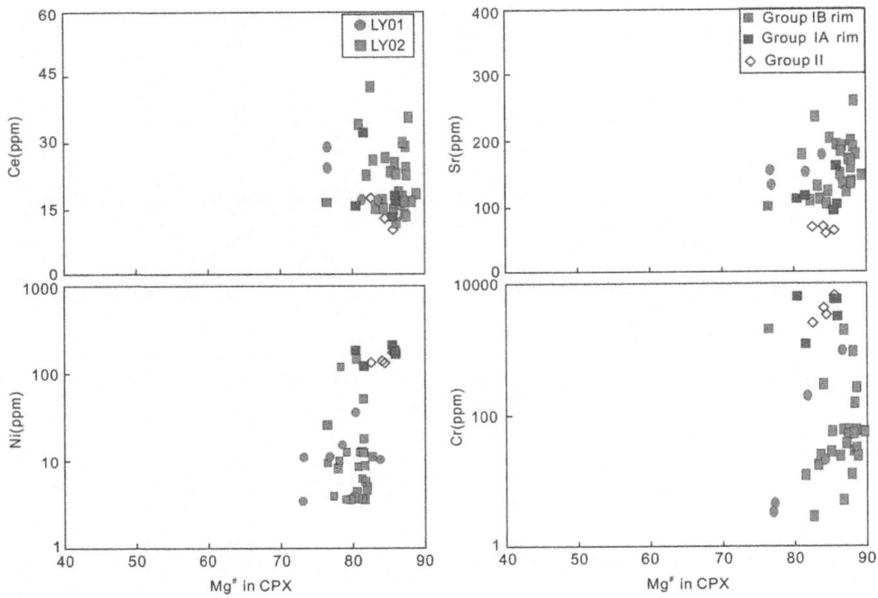

Fig. 3.19 Plots of Ce, Sr, Ni, and Cr versus Mg# for high-Mg clinopyroxene phenocrysts in the Jiaodong mafic dykes. *Data source* Group IA rim, Group IAB rim, Group II are from Liang et al. (2018); LY01 an LY02 are from Liang et al. (2019)

In the mantle source area of peridotite, clinopyroxene is the main carrier of incompatible elements; hence, the geochemical characteristics of clinopyroxene can effectively reveal the metasomatism of the source area (Rivalenti et al. 1996). Metasomatic melts and fluids in the mantle source may originate from the asthenospheric mantle (Xu et al. 2003; Zheng et al. 2007), circulating continental crust material (Gao et al. 2004, 2008), or subducting oceanic crust (Xu et al. 2013b; Zou et al. 2014). The high-Mg clinopyroxene phenocrysts in the Jiaodong mafic dyke show obvious Nb and Ta depletion (Fig. 3.20) and Sr isotope enrichment (Fig. 3.13), which eliminates the source of asthenospheric mantle (asthenosphere). The melts or fluids from the asthenosphere mantle source show no Nb and Ta depletion. Furthermore, the $^{87}Sr/^{88}Sr < 0.704$ is characteristic of MORBs (Zindler and Hart 1986), and $^{87}Sr/^{88}Sr < 0.708$ ratios are characteristic of OIBs (Ma et al. 2016). Moreover, continental crust (La/Nb = 2.50, U/Nb = 0.16; Rundnick and Gao 2003) and delamination of continental crust source melts have very low La/Nb and U/Nb ratios (Gao et al. 2008; Xu et al. 2013a), making it impossible to produce high La/Nb and U/Nb ratios of Jiaodong high Mg clinopyroxene (average La/Nb = 67.8, U/Nb = 0.36). The final possibility is that the melts and fluids may have originated from the subducting oceanic crust. High-Mg clinopyroxene shows high $^{87}Sr/^{86}Sr$ ratios, high Sr content, and a significant depletion in HFSEs (Figs. 3.11 and 3.12). These features are similar to the enriched mantle source of an island arc formed by subduction of the subducting oceanic crust

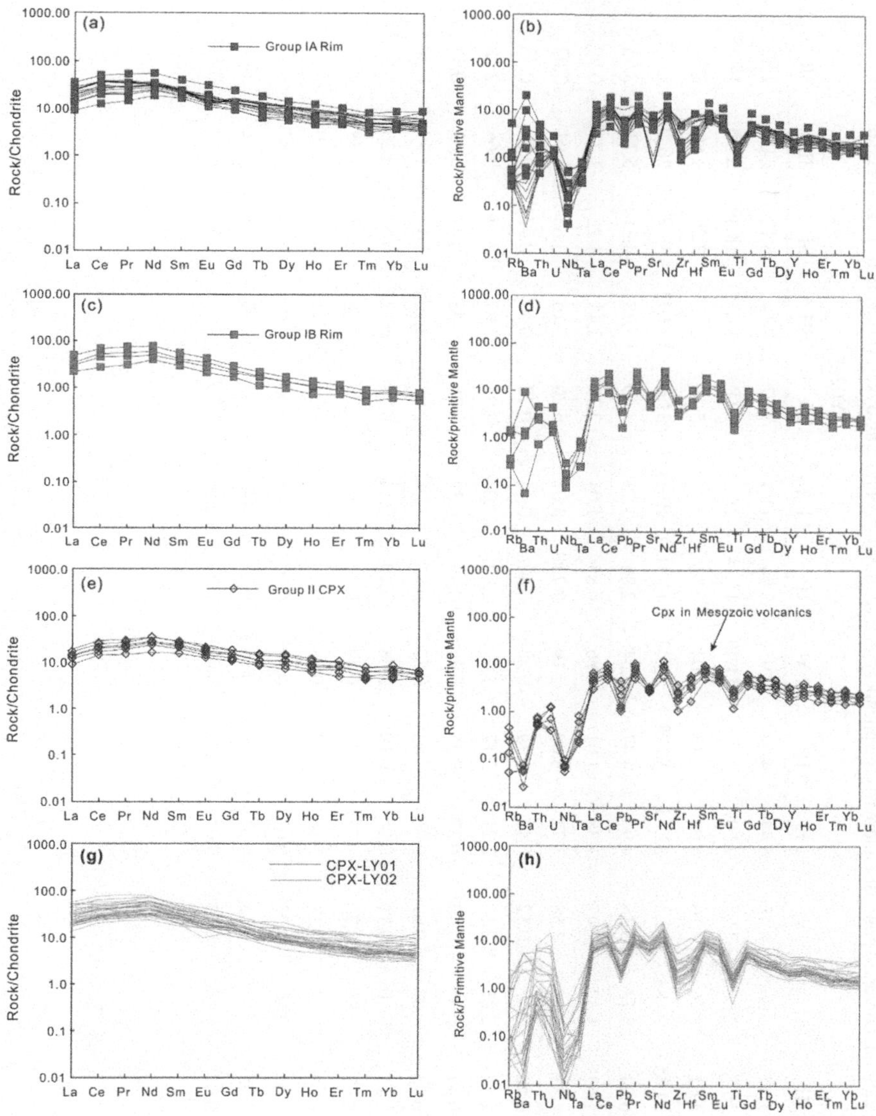

Fig. 3.20 **a** and **c** Chondrite-normalised REE patterns and **b, d** primitive mantle-normalised trace element patterns for high-Mg clinopyroxene phenocrysts in the Jiaodong mafic dykes. *Data source* Group IA rim, Group IAB rim, Group II are from Liang et al. (2018); LY01 an LY02 are from Liang et al. (2019)

Table 3.7 Representative trace element contents (ppm) of clinopyroxene phenocrysts from the Jiaodong mafic dykes. (Data are from Liang et al. 2019)

Sample	LY02-1										LY02-3				
Point	1C	1G	1H	1I	1 J	1 K	1L	1 M	1P	1–18	3A	3H	3I	3L	3R
La	6.02	12.23	7.27	6.06	6.48	6.44	4.42	2.95	3.94	11.44	7.67	4.14	5.17	7.06	4.22
Ce	16.78	42.62	26.23	22.60	23.56	22.91	17.16	11.48	13.18	28.85	29.83	14.71	18.49	25.70	14.65
Pr	2.90	7.11	4.62	3.96	4.10	4.08	3.13	2.04	2.32	3.93	5.28	2.62	3.29	4.60	2.32
Nd	16.35	34.91	24.35	20.83	22.36	22.15	16.98	10.42	13.31	19.23	31.61	14.92	17.97	26.36	13.47
Sm	4.24	6.73	6.04	5.13	5.36	4.97	4.47	2.54	2.89	3.77	7.60	3.78	4.31	5.73	3.05
Eu	1.00	2.00	1.62	1.32	1.47	1.38	1.01	0.86	0.85	1.04	2.37	1.11	1.22	1.77	0.99
Gd	3.21	4.77	4.33	3.94	3.58	3.77	3.00	2.19	2.73	3.01	6.20	3.16	3.68	5.04	3.03
Tb	0.39	0.59	0.48	0.47	0.52	0.48	0.39	0.28	0.40	0.44	0.78	0.40	0.47	0.59	0.36
Dy	2.47	2.66	2.52	2.21	2.30	2.21	2.17	1.31	2.02	2.08	4.95	2.04	2.14	3.66	2.28
Ho	0.47	0.52	0.45	0.35	0.40	0.37	0.35	0.26	0.43	0.39	0.76	0.42	0.41	0.63	0.38
Er	1.31	1.35	0.98	1.05	0.99	0.96	0.83	0.67	0.87	1.01	2.07	1.10	1.17	1.89	0.98
Tm	0.16	0.18	0.11	0.10	0.11	0.14	0.10	0.08	0.10	0.14	0.28	0.12	0.12	0.20	0.12
Yb	1.01	1.03	0.67	0.81	0.76	0.72	0.53	0.49	0.77	0.91	1.78	0.74	0.81	1.38	0.74
Lu	0.13	0.16	0.10	0.11	0.11	0.10	0.07	0.08	0.10	0.13	0.24	0.11	0.14	0.19	0.12
Rb	1.57	0.04	0.00	0.00	0.02	0.03	0.00	0.03	0.00	4.18	0.00	0.00	0.04	3.87	0.40
Ba	14.72	1.92	0.05	1.43	0.35	0.31	0.23	0.10	0.22	104.90	0.00	0.03	3.88	28.28	17.65
Th	0.22	0.12	0.07	0.08	0.03	0.04	0.03	0.02	0.04	1.39	0.10	0.04	0.06	0.25	0.13

(continued)

Table 3.7 (continued)

Sample	LY02-1										LY02-3				
Point	1C	1G	1H	1I	1 J	1 K	1L	1 M	1P	1–18	3A	3H	3I	3L	3R
U	0.04	0.03	0.01	0.01	0.01	0.00	0.00	0.01	0.02	0.09	0.01	0.00	0.01	0.02	0.02
Nb	0.15	0.04	0.01	0.02	0.01	0.01	0.02	0.01	0.01	0.69	0.07	0.02	0.03	0.60	0.06
Ta	0.00	0.00	0.00	0.00	0.00	0.00	0.00	0.00	0.00	0.02	0.02	0.00	0.00	0.02	0.00
Pb	0.36	0.21	0.21	0.35	0.12	0.16	0.18	0.15	0.15	1.82	0.14	0.11	0.18	0.44	0.28
Sr	127	236	203	187	196	191	169	145	132	158	135	122	131	130	110
Zr	27.5	37.3	26.1	20.4	18.5	22.8	11.0	7.0	13.4	37.8	61.4	16.6	25.6	45.7	23.7
Hf	1.05	1.43	1.29	0.69	0.80	1.01	0.47	0.28	0.45	1.58	2.74	0.78	1.17	2.05	0.93
Ti	2435	2091	2602	2226	2095	2415	1585	1156	1797	2327	4416	1956	2538	3490	1978
Y	11.14	13.64	10.12	9.51	10.15	9.98	8.58	6.71	9.06	10.02	21.33	9.67	10.37	17.91	10.61
V	142.2	124.2	95.4	90.1	88.0	92.6	62.2	105.6	129.5	216.1	186.7	98.6	120.9	155.9	189.4
Cr	4389	3064	4358	4464	4381	3384	2707	8213	7351	3601	326	3578	3892	3427	6149
Co	32.11	32.60	30.59	30.60	30.06	30.94	30.63	28.64	29.32	30.00	35.23	30.70	29.40	35.15	37.60
Ni	159	132	80	93	81	85	82	313	248	182	116	160	136	146	300

Sample	LY02-4						LY01-1					
Point	4D	4E	4F	4 K	4O	4–76	16G	16E	16 M	16L	16–70	16–71
La	4.22	3.70	6.46	4.76	4.20	10.23	4.72	8.46	5.56	5.05	6.16	4.45
Ce	16.34	13.48	22.87	18.09	16.31	35.87	15.95	28.95	22.03	16.84	23.97	16.54
Pr	3.21	2.38	3.85	3.04	3.10	6.18	2.57	4.98	3.96	2.75	4.20	2.99
Nd	17.88	14.07	20.89	16.45	18.22	31.31	13.33	26.51	21.26	15.08	24.52	17.29
Sm	4.40	3.51	4.64	3.99	4.72	7.16	3.71	6.09	4.76	4.26	5.48	4.11

(continued)

Table 3.7 (continued)

Sample	LY02-4						LY01-1					
Point	4D	4E	4F	4 K	4O	4–76	16G	16E	16 M	16L	16–70	16–71
Eu	1.25	0.92	1.36	1.08	1.39	2.02	1.02	1.95	1.50	1.04	1.85	1.15
Gd	3.52	2.84	3.91	3.24	3.59	5.23	2.89	5.03	3.68	3.44	4.53	3.10
Tb	0.34	0.37	0.45	0.37	0.40	0.61	0.40	0.74	0.48	0.35	0.60	0.36
Dy	1.79	2.10	2.44	2.13	1.98	2.79	2.17	3.99	2.81	2.08	3.93	1.88
Ho	0.32	0.36	0.40	0.32	0.36	0.48	0.37	0.73	0.45	0.41	0.70	0.33
Er	0.63	0.90	1.01	0.87	0.74	1.01	1.06	1.88	1.32	1.12	1.74	0.73
Tm	0.09	0.11	0.14	0.14	0.11	0.12	0.11	0.28	0.20	0.13	0.18	0.11
Yb	0.59	0.71	0.72	0.85	0.58	0.79	0.79	1.60	0.83	0.83	1.27	0.77
Lu	0.08	0.09	0.11	0.09	0.09	0.11	0.10	0.31	0.15	0.12	0.17	0.09
Rb	0.02	0.05	0.00	0.02	0.09	0.01	0.62	1.12	0.07	0.93	0.00	0.00
Ba	2.12	0.85	0.10	4.51	0.17	0.22	40.88	25.72	0.14	39.55	0.00	0.18
Th	0.02	0.03	0.06	0.06	0.04	0.08	0.18	0.38	0.05	0.24	0.06	0.03
U	0.03	0.01	0.01	0.00	0.00	0.00	0.04	0.06	0.00	0.04	0.01	0.00
Nb	0.02	0.00	0.01	0.03	0.02	0.01	0.08	0.10	0.03	0.12	0.05	0.00
Ta	0.00	0.00	0.00	0.00	0.01	0.00	0.00	0.02	0.00	0.00	0.01	0.00
Pb	0.21	0.14	0.18	0.09	0.16	0.18	0.40	0.35	0.13	0.44	0.12	0.16
Sr	190	133	169	146	178	259	143	154	139	150	130	177
Zr	8.0	14.0	14.9	14.9	8.6	16.0	19.7	44.1	25.9	20.3	36.6	10.0
Hf	0.33	0.56	0.60	0.64	0.46	0.73	0.78	1.83	1.12	0.87	1.63	0.52
Ti	1177	1840	1403	1683	1185	1756	1888	3787	2688	2127	2975	1453

(continued)

Table 3.7 (continued)

Sample	LY02-4						LY01-1					
Point	4D	4E	4F	4 K	4O	4–76	16G	16E	16 M	16L	16–70	16–71
Y	7.33	8.75	10.01	9.51	9.05	11.53	9.61	19.12	12.78	10.38	15.99	8.62
V	52.7	91.3	83.4	91.5	54.0	80.0	166.9	199.7	133.7	119.3	138.0	126.0
Cr	3771	4286	5451	4378	3406	6077	7438	1245	1886	5704	1577	3243
Co	29.21	28.62	29.46	29.04	28.91	27.24	30.03	32.27	33.09	29.69	34.48	32.91
Ni	110	160	137	153	101	97	226	78	147	151	150	171

(Hofmann 1988; Tatsumi and Takeshi 2003). The melt from the oceanic crust is characterised by enriched Sr isotopes and shows the characteristics of adakites (Hermann and Rubatto 2009). However, the high-Mg clinopyroxene in Jiaodong does not show any similarities to adakites, so the source area of the oceanic crust melts can be excluded as well. In contrast, the water-containing fluid formed by the dehydration of subducting oceanic slabs has a high Sr content and $^{87}Sr/^{86}Sr$ ratios (Kogiso et al. 1997). Moreover, HFSEs have low mobility in water-containing fluids (Kogiso et al. 1997; Zou et al. 2014). As a result, the lithospheric mantle has high Sr contents and $^{87}Sr/^{86}Sr$ ratios but shows depletion in HFSEs after metasomatism by water-bearing fluids derived from subducting oceanic crust. This is consistent with the geochemical characteristics of the high-Mg clinopyroxene in Jiaodong mafic dykes (Figs. 3.12, 3.20). Moreover, Nb has lower mobility than La and U in aqueous fluids (Kogiso et al. 1997; Xu R. et al. 2013), so subduction-related fluids should have high La/Nb and U/Nb ratios, which has been observed in the Jiaodong high-Mg clinopyroxene (Fig. 3.20). The $^{87}Sr/^{86}Sr$ ratios ($^{87}Sr/^{86}Sr = 0.7081–0.7088$) of high-Mg clinopyroxene and host mafic dyke in Jiaodong is higher than that of the plate-derievd fluids ($^{87}Sr/^{86}Sr = 0.7041$, Sr = 13530 ppm; Tatsumi 2000). However, the seawater altered oceanic crust ($^{87}Sr/^{86}Sr = 0.7092$; Tatsumi 2000) or the oceanic sediments ($^{87}Sr/^{86}Sr = 0.7053–0.7312$; Plank and Langmuir 1998) may have been suitable metasomatic media of the source. Nevertheless, the $\varepsilon_{Nd}(t)$ value (MORB > 0; Hofmann 2003) of seawater altered oceanic crust is much higher than that of the Jiaodong mafic dyke (from −15.6 to −12.6) (Fig. 3.5). Therefore, the oceanic crust altered by seawater could not have been the only metasomatic medium as it cannot explain the extremely low $\varepsilon_{Nd}(t)$ value of the Jiaodong mafic dyke. Tatsumi and Takeshi (2003) found that the addition of fluids from the slab in the subduction zone does not greatly affect the Sr–Nd–Pb–Hf isotope composition of the source area, but the addition of subducted sediments and altered oceanic crust can effectively change the composition of mantle-derived isotopes (Fig. 3.5) (Spandler and Pirard 2013; Dai et al. 2017; Khan et al. 2017; Zhang et al. 2017). Therefore, the Jiaodong mafic dyke and high-Mg clinopyroxene may have originated from an enriched lithospheric mantle metasomatised by subducted oceanic crust altered by seawater, which contains oceanic sediments.

3.5.2 High H_2O Contents of Clinopyroxene Phenocrysts

3.5.2.1 H_2O Contents of Clinopyroxene Phenocrysts

The H_2O content of clinopyroxene phenocrysts was determined by infrared spectroscopy at the School of Earth Sciences, Zhejiang University. The sample was polished on both sides, and the thickness of the disc was approximately 130–180 μm. The polished sheet was retracted from the glass sheet, soaked in acetone and alcohol to remove the resin from the surface, and then directly analysed under the infrared laser. For samples with relatively low water content, the infrared spectroscopy instrument, including light source, light path, and sample chamber need to be purged with dry gas (high-purity N_2 or dry air that removes moisture and CO_2), to minimise the interference of water vapour in the air. In the infrared spectrum analysis, the centre of the mineral grain is generally selected. The spot size mainly depended on the size of pyroxene grains, which are generally 30 μm × 30 μm to 100 μm × 100 μm in size. The scanning times of the background and samples are generally 126–256 times, depending on the water content of pyroxene and the dryness of air in the laboratory. For large clinopyroxene grains, a cross-sectional analysis from core to rim was acquired to check whether there is any loss of water by diffusion. After baseline correction of the obtained spectrum, the Beer-Lambert law was used to calculate the water content (Xia et al. 2013).

The basis for the quantitative analysis of water content in minerals by infrared spectroscopy is the Beer-Lambert law. The formula is expressed as $\boldsymbol{c = A/(I \times t)}$, where $\boldsymbol{A}$ is the total integrated absorption intensity (that is, the integrated area of the absorption peak, in cm^{-1}), $\boldsymbol{t}$ is the thickness of the analysed mineral grain or disk (cm), $\boldsymbol{I}$ is the infrared absorption coefficient of water in the mineral (for clinopyroxene (REF) $10^{-6}/cm^2$), and $\boldsymbol{c}$ is the obtained water content in the mineral. The absorption coefficient I depends on the type of mineral and the position of the OH absorption peak. For the common diopside pyroxene phenocrysts in basalt, Bell et al. (1995) obtained the absorption coefficient (7.84 × 10–6/cm^2) based on pyroxene from the South African Kimberlite pipe. The value was corrected by the widely-used hydrogen pressure gauge method to independently measure the water content of the ordinary pyroxene crystal. The thickness (t) was obtained directly using a spiral micrometre. The OH absorption peak of clinopyroxene is between 3000 and 3700 cm^{-1} (Kovács et al. 2008), and the A value is three times the integral area. The average water content of clinopyroxene in each sample was used to determine the water content of the source melt. The water contents of the clinopyroxene phenocrysts in the Jiaodong mafic dyke are shown in Table 3.8. The absorption peak of clinopyroxene phenocrysts is between 2800 and 3700 cm^{-1}, which is consistent with the results of earlier studies (Xia et al. 2013; Doucet et al. 2014; Liu et al. 2015b) (Fig. 3.21a).

O'Leary et al. (2010) experimentally measured the partition coefficient of water between clinopyroxene and basaltic melts containing variable amount of Al. Combined with the previous studies (Wade et al. 2008), partition coefficient data were used to re-fit the relationship between the content of four-coordinated Al and M-site Ca in the Dcpx/melt H_2O and clinopyroxene. The relationship can be described as $\ln D^{cpx/melt}_{(H2O)} = -4.2\ (\pm\ 0.2) + 6.5\ (\pm\ 0.5) \times \text{cpx Al}^{IV} - 1.0\ (\pm\ 0.2) \times \text{cpx Ca}$. In this study, clinopyroxene phenocrysts with Mg# > 85 were selected to calculate the water content of the magma. The calculation results showed that the water content of clinopyroxene was 123–574 ppm. According to the distribution coefficient $D^{cpx/melt}_{(H2O)}$ of water between clinopyroxene phenocrysts and the melt (O'Leary et al. 2010), the calculated water content of the melt in equilibrium with clinopyroxene is 1.38–4.45 wt% (Fig. 3.21b). The Mg# value of the equilibrium with melt of the high-Mg clinopyroxene phenocryst (including Mg# > 85) in the mafic dyke rocks in Jiaodong is consistent with the highest Mg# value of the whole rock (Fig. 3.21c). Therefore,

Table 3.8 EMP and FTIR analytical results for clinopyroxene phenocrysts from the Jiaodong mafic dykes. (Data are from Liang et al. 2019)

Sample	Point	SiO_2	TiO_2	Al_2O_3	^{T}FeO	MnO	MgO	CaO	Na_2O	Cr_2O_3	P_2O_5	Total	$Mg^{\#}$	^{IV}Al	D	IR integrated area (cm^{-1})	Thickness (cm)	H_2O (cpx, ppm)	H_2O (Melt, wt%)
LY02-1	1A	52.83	0.11	2.17	5.65	0.5	14.75	21.65	1.09			98.75	82.31	0.0318	0.0078				
	1C	51.62	0.37	3.35	4.5	0.32	17.67	19.15	0.7	1.18		98.86	87.50	0.0963	0.0132				
	1G	53.32	0.14	2.04	5.61	0.36	15.61	21.11	0.98	0	0	99.17	83.22	0.0287	0.0078				
	1H	51.66	0.43	2.43	5.31	0.17	17.18	21.38	0.34	0.65		99.55	85.22	0.0922	0.0117	10.91	0.0144	321	2.74
	1I	52.08	0.39	2.16	4.65	0.15	17.14	20.87	0.39	0.77	0.03	98.63	86.79	0.0686	0.0102	12.90	0.0144	379	3.71
	1J	53.04	0.25	1.32	4.34	0.25	18.12	20.78	0.35	0.51		98.96	88.15	0.0458	0.0089	4.18	0.0144	123	1.38
	1K	51.91	0.43	2.31	4.93	0.19	17.23	21.07	0.31	0.65	0.07	99.1	86.17	0.0794	0.0109	10.54	0.0144	310	2.84
	1L	53.12	0.25	1.35	4.38	0.06	17.74	21.26	0.23	0.53	0.04	98.96	87.83	0.0422	0.0085	12.89	0.0144	379	4.45
	1M	53.17	0.27	1.37	4.87	0.14	17.93	20.96	0.28	0.47		99.46	86.78	0.0473	0.0089	9.36	0.0144	275	3.08
	1N	50.48	0.86	4.71	6.13	0.17	15.75	20.59	0.52	0.55		99.76	82.08	0.1367	0.0161				
	1O	53.57	0.13	2.3	4.54	0.18	16.58	20.99	0.96	0.4		99.65	86.68	0.0397	0.0085				
	1P	52.32	0.31	3.32	4.33	0.15	17.85	19.35	0.64	1.17	0	99.44	88.02	0.0867	0.0123	10.68	0.0139	325	2.63
	1R	52.88	0.37	2.15	4.91	0.16	18.25	19.7	0.46	0.34	0.07	99.29	86.89	0.0597	0.0102				
	1S	50.39	0.75	4	6.3	0.14	15.67	20.7	0.42	0.38		98.75	81.60	0.1194	0.0142				
	1–18	52.95	0.31	1.98	4.34	0.15	18.06	20.66	0.35	0.72	0.02	99.54	88.12	0.0613	0.0099				
LY02-3	3A	52.53	0.36	3.11	4.48	0.19	18.06	19.86	0.55	1.11	0.03	100.28	87.78	0.0905	0.0125				
	3B	51.08	0.52	3.57	5.28	0.11	16.74	20.73	0.43	0.95	0.01	99.42	84.96	0.1136	0.0138				
	3G	51.99	0.35	2.7	4.49	0.16	17.4	20.74	0.43	1.01		99.27	87.35	0.0856	0.0115				
	3H	52.42	0.35	3.34	4.7	0.18	17.83	19.3	0.68	0.99		99.79	87.12	0.0874	0.0124				

(continued)

Table 3.8 (continued)

Sample	Point	SiO_2	TiO_2	Al_2O_3	^{T}FeO	MnO	MgO	CaO	Na_2O	Cr_2O_3	P_2O_5	Total	$Mg^{\#}$	^{IV}Al	D	IR integrated area (cm^{-1})	Thickness (cm)	H_2O (cpx, ppm)	H_2O (Melt, wt%)
	3I	50.7	0.69	4.5	6.11	0.15	16.22	20.74	0.41	0.69		100.21	82.55	0.1366	0.0161				
	3L	50.33	0.83	4.92	6.43	0.22	15.87	20.95	0.43	0.26	0.01	100.25	81.48	0.1474	0.0171	24.69	0.0182	574	3.36
	3 M	52.47	0.48	2.45	6.11	0.15	17.43	20.47	0.33	0.35	0.04	100.28	83.57	0.0790	0.0112				
	3R	52.21	0.45	3.82	4.43	0.07	15.91	22.75	0.41		0.02	100.07	86.49	0.0911	0.0111				
	3S	53.64	0.32	1.4	5.63	0.2	18.36	20.33	0.26	0.15	0.03	100.32	85.32	0.0459	0.0091				
	3Q	51.06	0.73	3.87	5.69	0.24	16.65	20.62	0.37	0.7		99.93	83.91	0.1219	0.0147				
	3–66	47.02	0.56	10.84	4.24		13.28	22.63	0.42	0.05		99.04	84.81	0.2596	0.0330				
	3–67	53.32	0.1	1.91	6.98	0.14	14.9	21.95	0.72	0.08		100.1	79.19	0.0330	0.0078				
LY02-4	4D	52.14	0.4	2.18	4.17	0.11	17.85	20.74	0.36	1.18		99.13	88.41	0.0795	0.0111				
	4E	52.33	0.33	2.3	4.26	0.19	17.67	21.23	0.48	0.92	0.08	99.79	88.09	0.0807	0.0110				
	4F	53.98	0.21	1.09	4.51	0.11	18.97	19.92	0.48	0.72		99.99	88.23	0.0372	0.0088	11.59	0.0142	345	3.93
	4K	53.7	0.19	0.93	3.96	0.18	19.11	20.04	0.31	0.59		99.01	89.59	0.0324	0.0084				
	4M	53.75	0.18	1.02	3.96	0.25	18.87	20.46	0.35	0.6	0.07	99.51	89.47	0.0359	0.0085				
	4N	52.99	0.49	2.11	5.19	0.16	17.79	20.99	0.43	0.32	0.02	100.49	85.93	0.0699	0.0104				
	4U	53.3	0.21	1.3	5.13	0.26	18.03	20.66	0.31	0.27		99.47	86.23	0.0424	0.0088				
	4V	52.79	0.14	3.25	3.97	0.07	15.6	22.6	0.74	0.05		99.21	87.51	0.0588	0.0090				
	4O	52.67	0.33	2.06	3.98	0.16	17.55	21.16	0.43	1.08	0.01	99.43	88.71	0.0669	0.0101				
	4–76	52.51	0.26	2.18	4.05	0.17	17.65	20.95	0.45	1.07	0	99.29	88.59	0.0702	0.0104				

(continued)

Table 3.8 (continued)

Sample	Point	SiO_2	TiO_2	Al_2O_3	^{T}FeO	MnO	MgO	CaO	Na_2O	Cr_2O_3	P_2O_5	Total	$Mg^{\#}$	^{IV}Al	D	IR integrated area (cm^{-1})	Thickness (cm)	H_2O (cpx, ppm)	H_2O (Melt, wt%)
LY01	6G	53.23	0.21	2.42	5	0.09	18.19	19.64	0.5	0.49	0.01	99.78	86.64	0.0589	0.0102	11.56	0.0149	328	3.22
	6E	48.05	1.25	5.96	7.64	0.09	14.34	20.93	0.46	0.05	0.01	98.78	76.99	0.1901	0.0222				
	6 M	53.6	0.21	1.67	3.13	0.04	17.65	22.4	0.3	0.08		99.08	90.95	0.0368	0.0079				
	6L	51.83	0.58	3.04	6.52		16.5	20.46	0.42	0.17		99.52	81.85	0.0861	0.0117				
	6–70	52.61	0.1	2.05	7.29	0.23	13.8	22.79	0.51	0		99.38	77.14	0.0366	0.0076				
	6–71	52.16	0.51	2.39	5.79	0.24	17.25	20.35	0.41	0.22		99.32	84.15	0.0745	0.0109				

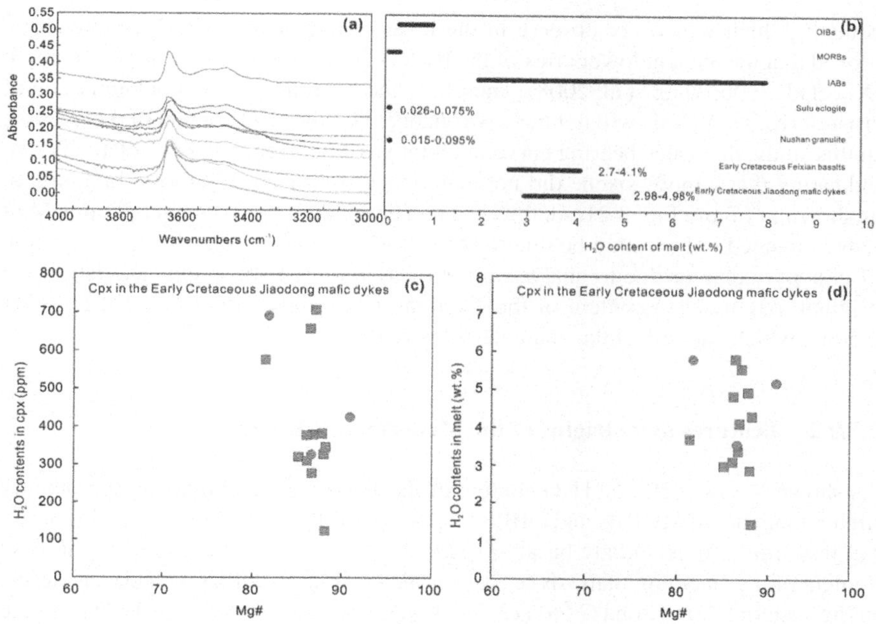

Fig. 3.21 **a** Representative unpolarised FTIR spectra of clinopyroxene phenocrysts; **b** Comparison of H_2O content of the primary melts of the Jiaodong mafic dykes and the ranges of H_2O contents for OIB, MORB, IAB, Sulu eclogite, Nushan granulite, and Feixian basalts; H_2O contents of **c** clinopyroxene phenocrysts and **d** melts versus Mg# of clinopyroxene phenocrysts in the Jiaodong mafic dykes. *Data source* Comparative data were taken from Dixon et al. (2004), Xia et al. (2006, 2013), Sheng et al. (2007), and Chen et al. (2015); Jiaodong Early Cretaceous mafic dykes are from Liang et al. (2019)

these clinopyroxene phenocrysts are the earliest minerals crystallised from the original basalt melt and are suitable for calculation of the H_2O content in the original magma.

In this study, we selected clinopyroxene with high Mg# (>85) as the most suitable grains for determining their water content. We then used average H_2O content of each sample for calculating the H_2O content in the equilibrium melt, based on the distribution coefficient $D_{(H2O)}^{cpx/melt}$ (Liu et al. 2015a, b). We obtained the H_2O content in the original magma to be 1.38–4.45 wt% (Fig. 3.21b). The calculation of the H_2O content in basaltic melt requires comprehensive analysis of factors such as separation by crystallisation, degassing, crustal assimilation, contamination, and loss of H_2O content by diffusion in clinopyroxene phenocrysts after crystallisation (Woods et al. 2000; Liu et al. 2015b; Liu et al. 2016). As shown in Fig. 3.21c-d, with the decrease in Mg#, the H_2O content in clinopyroxene does not increase significantly, indicating that the effect of crystallisation on the H_2O content is not obvious. The influence of magma degassing can also be ruled out because as Mg# decreases, the H_2O content in the melt does not decrease. At the same time, degassing causes a variable H_2O

content, which we do not observe in the investigated samples. Early studies have proved that the ancient lower crust of the North China Craton has a low H_2O content (Xia et al. 2006; Yang et al. 2008). Therefore, the assimilation and contamination of crust-derived materials will not increase but decrease the H_2O content of the magma. In this study, the water-bearing curve in clinopyroxene phenocrysts is relatively flat, indicating that clinopyroxene did not experience significant H_2O loss. In addition, as shown in Figure Fig. 3.21c-d, the H_2O content of clinopyroxene in the same Mg# range changed, which may be due to the diffusion of H_2O after the crystallisation of clinopyroxene or due to analysis errors. Based on the above characteristics, the original magma H_2O content of the Jiaodong mafic dyke is not affected by other factors, which can reflect the characteristics of its source.

3.5.2.2 Features and Origins of the Metasomatic Agents

As shown in Fig. 3.20, the H_2O content of the Jiaodong mafic dyke is significantly higher than that of MORBs and OIBs (MORB = 0.1–0.3 wt%; OIB = 0.3–1.0 wt%), but it is similar to island arc basalts (IABs; 2.7–4.1 wt%; Wallace 2005). The H_2O content of the Jiaodong mafic dyke rocks is also similar to that of the Feixian basalts in the eastern North China Craton (2.7–4.1 wt%), which sourced their H_2O from the mantle area that was exposed to multiple oceans before the craton was thinned, and the plate subduction contributed to enrichment (Xia et al. 2013). These characteristics further indicate that the Jiaodong mafic dyke is derived from the lithospheric mantle with high H_2O content, rather than from a MORB or OIB. H_2O and Ce have similar incompatibilities in olivine, clinopyroxene, and plagioclase (Dixon et al. 2002), so the H_2O/Ce ratio is relatively stable during partial melting and separation crystallisation. Based on this, the H_2O/Ce ratio of basalt magma can be used to indicate the nature of its mantle source (Wade et al. 2008; Xia et al. 2013). A large number of studies have proved that, based on the Ce of the whole rock, the H_2O/Ce ratio of primary magma will not change significantly during the partial melting and separation crystallisation process (Liu et al. 2015b). This study combines the H_2O and Ce content of the whole rock samples traces the properties of the mantle source area through the H_2O/Ce ratio.

(1) **Mantle enrichment medium: continental crustal material that is not subduction cycle or delamination**

Although the lower crustal delamination of the North China Craton or subduction of the Yangtze plate material can explain the whole-rock geochemical characteristics of the Jiaodong basic dyke, they cannot explain the high H_2O content of the original magma of the dyke. Based on the H_2O content of anhydrous clinopyroxene (~0.236 wt%), orthopyroxene (~0.117 wt%), and plagioclase (~0.088 wt%) (Xia et al. 2006), the H_2O content of granulite enclaves in the Cenozoic Nushan basalt in the eastern part of the North China Craton is determined to be 0.015–0.095 wt% (Fig. 3.20b). The average Ce content of granulite in the lower crust of Nushan is 112 ppm (Huang et al. 2004). Based on the above data, the H_2O/Ce values of the lower continental crust of the eastern North China Craton are 1.34–8.50, which is significantly lower than that of the Jiaodong mafic dyke rocks (237–322). Therefore, it can be ruled out that the lower crustal delamination of the North China Craton contributed the high H_2O/Ce ratio of the Jiaodong mafic dyke. During the Triassic period, the Yangtze Plate subducted to the southeast of the North China Craton, and the Neoproterozoic basalts underwent Ultra-High Pressure metamorphism, forming a large amount of eclogites in the Dabie orogenic belt (Zheng 2012). During continental subduction, eclogite can bring hundreds of ppm of water into the mantle (Xia et al. 2005). Based on the calculation of the water content of garnet and pyroxene, the H_2O content of the above-mentioned eclogite is 0.026–0.075 wt% (Fig. 3.20b; Sheng et al. 2007). The average Ce content of the UHP metamorphic eclogite in the Dabie orogenic belt is 16 ppm (Zhao et al. 2007). It is calculated that the H_2O/Ce ratio of the eclogite in the eastern part of the North China Craton is 16–46, which is significantly lower than that of the Jiaodong mafic dyke. We, therefore, conclude the high H_2O/Ce characteristics of Jiaodong mafic dyke rocks are not derived from the subduction of the Yangtze plate.

(2) **Mantle enrichment medium: subducting oceanic slab and marine sediment source fluid**

The high H_2O content of the Jiaodong mafic dyke rocks may originate from the oceanic crust, because in the subduction zone, marine sediments and water-rich oceanic crust have relatively high-water content (Alt and Teagle 2003; Kendrick et al. 2013). It is worth noting that the H_2O/Ce ratios of Jiaodong mafic dyke rocks are significantly higher than that of normal oceanic crust (such as MORB $\leq$ 200) (Fig. 3.22; Asimow et al. 2004). However, the H_2O/Ce ratios of seawater-altered MORB and marine sediments can be as high as 5000 and 1200, respectively (Dixon et al. 2002). From this, we speculate that the oceanic slabs altered by seawater and marine sediments could have contributed the high H_2O contents in the mantle source of the Jiaodong mafic dyke rocks. As shown in Fig. 3.22a–b, the fluid metasomatism produced by subducting oceanic crust alone cannot produce such high H_2O/Ce and $(Ba/Th)_N$ and low $(Nb/La)_N$ values, indicating that there are still other contributing factors. Marine sediments may be one of the media, because the marine sediment carried by the subducting oceanic crust can significantly increase the water content in the mantle (Windley et al. 2010; Liu et al. 2015b), and at the same time, the $(Ba/Th)_N$ and $(Nb/La)_N$ values (Martin et al. 2005; Liu et al. 2015a). In addition, the whole-rock geochemical characteristics of the mafic dyke rocks can further prove that the mantle source area is affected by marine sediments. The ratios of some incompatible elements (such as Ba/La, Nb/La, Ce/Pb, and Nb/U) indicative of metasomatism in the mantle source (Prelević et al. 2012). The Jiaodong mafic dyke rocks have higher Ba/La and lower Nb/U, Nd/La, and Ce/Pb ratios (Fig. 3.22c–d), which are consistent with the characteristics of marine sediments. The addition of altered oceanic crust and marine sediments can also explain the Sr–Nd isotopic characteristics of the

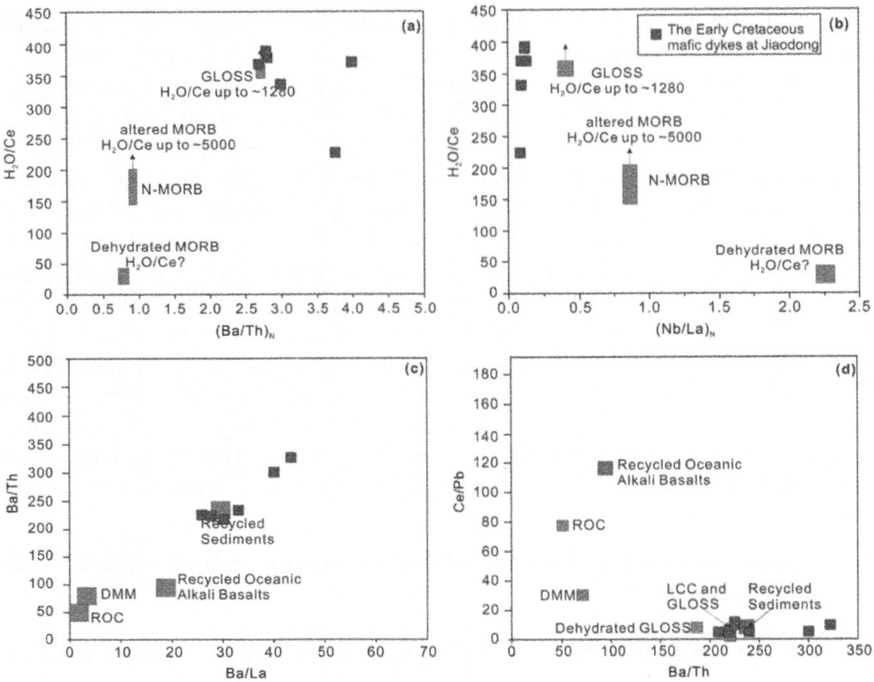

Fig. 3.22 Plots of **a** H_2O/Ce versus $(Ba/Th)_N$; **b** H_2O/Ce versus $(Nb/La)_N$; **c** Ba/Th versus Ba/La; **d** Ce/Pb versus Ba/Th (After Liang et al. 2019). Data source: H_2O/Ce ratios of N-MORB, altered MORB, and GLOSS are from Michael (1995) and Dixon et al. (2002); trace element data of N-MORB, altered MORB, GLOSS; recycled sediments are from Plank and Langmuir (1998), Workman and Hart (2005), and Gale et al. (2013) and the Early Cretaceous mafic dykes at Jiaodong are from Liang et al. (2019)

whole dyke rock (Fig. 3.5). In summary, the water content, and geochemical characteristics of the Jiaodong mafic dyke rock indicate that its mantle source area has been metasomatised by subducting oceanic slab source fluids and marine sediments.

3.6 Lithospheric Mantle Was Changed During Mesozoic

3.6.1 Nature of the Plaleoproterozoic Lithospheric Mantle

The Paleoproterozoic mafic dyke rocks in the Qixia area of the Jiaobei terrane follow the northeast-southwest direction and are approximately 30 cm thick. The main mineral assemblages are plagioclase and amphibole, which have undergone amphibolite-facies metamorphism. The zircon grains in the Paleoproterozoic mafic dyke selected for dating are light grey–black, with an euhedral to semi-euhedral prismatic structure, and showing oscillatory zoning, have high Th (49–589 ppm) and U (86–510 ppm) content, and high Th/U ratios (0.52–1.16), and show a positive Ce anomaly and enrichment in HREE in the chondrite-standardised REE diagram. These characteristics indicate that zircon is crystallised from magma and not formed or affected during metamorphism (Corfu et al. 2003; Zeng et al. 2018). The $^{207}Pb/^{206}Pb$ weighted average age of these zircons is 2158 $\pm$ 24 Ma (MSWD = 0.85, n = 10) (Liang et al. 2020b). This crystallisation age is at the same as that of the Paleoproterozoic metamorphic volcanic rocks and granites exposed in the Jiaoliao-Jiji ancient rift zone (JLJB). For example, the Laiyang metamorphic gabbro of similar (2102 $\pm$ 3 Ma) has been documented in the Jiaobei terrane in southern JLJB (Liu et al. 2013a). Further rocks of similar ages have been reported: Changyi alkali feldspar granite (2193 Ma) and albite granite(2171 Ma; Lan et al. 2015) and Qixia granite (2181 $\pm$ 12 Ma; Liu et al. 2014); Sea Urban mafic dykes of North JLJB (2115–2130 Ma; Wang et al. 2016c; Xu et al. 2018; Yuan et al. 2015); metamorphic volcanic rocks in the Ji'an Group (2175–2240 Ma, Wang et al. 2018); metamorphic rhyolite and metamorphic dacite in the Nanliaohe Group (2158–2205 Ma; Li and Chen 2014; Li et al. 2018); metamorphic rhyolite and metamorphic dacite in the Beiliaohe Group (2184–2127 Ma; Bi et al. 2018); the Liaoji granite in the Nanliaohe Group (2166–2176 Ma, Li and Zhao 2007; 2171–2185 Ma, Liu et al. 2018); metamorphic volcanic rocks in the Wuhu Group at the southern end of the JLJB (2126 $\pm$ 37 Ma; Liu et al. 2018b).

The Paleoproterozoic mafic dyke rock in Jiaobei has low SiO_2 (49.7–55.3 wt%) and $Na_2O + K_2O$ (1.80–3.71 wt%) contents, as well as high contents of Al_2O_3 (12.0–15.2 wt%). MgO (6.57–9.49 wt%), Cr (44–286 ppm) and Ni (36–103 ppm) (Table 3.9; Liang et al. 2020b). In the Zr/TiO_2-Nb/Y chart, the mafic dyke data falls on the basalt series (Fig. 3.23). The Jiaobei mafic dyke has a high MgO content ($\leq$9.29 wt%) and high Mg# value ($\leq$62), indicating its mantle origin (Frey et al. 1978). Compared with the MORB or OIB magma derived from the asthenospheric mantle, the Jiaobei mafic dyke shows different trace element characteristics, such as Nb and Ta depletion (Fig. 3.24). The Jiaobei mafic dyke shows a weak Eu anomaly (δEu = 0.60–1.15) (Fig. 3.24a) and significant HFSE depletion characteristics (Fig. 3.24b), with a value of $\varepsilon_{Nd}(t)$ from -2.07 to -0.09, and the Nd model age of 2722–2967 Ma (Fig. 3.25). These features are similar to those of Paleoproterozoic mafic dykes derived from the continental lithospheric mantle of the North China Craton (Peng et al. 2008, 2012). The zircon in the Jiaobei mafic dyke has a low $^{176}Lu/^{177}Hf$ ratio (<0.004), indicating that radioactive Hf was not enriched after the zircon crystallised, indicating that the $^{176}Hf/^{177}Hf$ shows the primordial value (Cheng et al. 2017). The Jiaobei mafic dyke also shows high radioactive Lu–Hf isotope values (εHf(t) = -13.6 to -4.0), which is similar to the basic rocks derived from the lithospheric mantle of the North China Craton in the Paleoproterozoic. Moreover, the Nd and Hf isotope model ages (TDM1) are 2722–2896 Ma and 2627–2870 Ma, respectively, consistent with the Archean crustal accretion period in the eastern North China Craton (Liu et al. 2013c; Peng et al. 2012). Therefore, the Paleoproterozoic mafic dykes may originate from the lithospheric mantle of the North China Craton. In addition, these mafic dykes have island arc characteristics, such as depletion in HFSEs (for example, Nd and

Table 3.9 Whole-rock geochemical compositions of the Jiaobei mafic dykes. (Data are from Liang et al. 2020c)

Sample	QX-1		QX-2			QX-3		Standards	Detection limit
	QX-1-1	QX-1-2	QX-2-1	QX-2-2	QX-2-3	QX-3	QX-3	BHVO-2	
SiO_2	53.77	52.89	49.69	52.02	50.13	54.77	55.15	49.72	0.1
TiO_2	0.31	0.38	0.72	0.68	0.72	0.88	0.86	2.76	0.05
Al_2O_3	13.00	13.07	12.58	12.03	12.56	12.21	12.05	13.44	0.1
TFe_2O_3	10.08	10.63	13.97	13.06	13.77	12.25	12.12	12.4	0.1
MnO	0.15	0.17	0.19	0.21	0.20	0.17	0.18	0.17	0.05
MgO	8.80	9.02	9.49	8.37	8.80	7.32	7.29	7.23	0.05
CaO	9.30	9.27	9.85	9.54	9.95	8.72	8.87	11.24	0.05
Na_2O	2.59	2.67	2.77	2.73	2.81	2.42	2.33	2.22	0.05
K_2O	1.03	1.01	0.83	0.98	0.86	0.66	0.63	0.52	0.05
P_2O_5	0.04	0.04	0.08	0.07	0.07	0.14	0.13	0.27	0.05
LOI	1.61	1.52	0.75	1.13	1.08	1.03	0.96	0	0.1
Total	99.83	99.82	99.82	99.79	99.79	99.79	99.77	99.97	
Mg#	61	60	55	53	53	52	52		
Li	16.26	15.97	17.02	14.97	19.81	7.31	7.65	4.7	1
Be	0.24	0.27	0.47	1.13	0.35	0.51	0.49	1.06	0.2
Sc	42.06	46.29	52.1	50.13	51.71	42.34	43.15	31.88	1
V	189.3	230.5	312.4	279.4	285.1	264.7	264	323	5
Cr	95	81.8	124.6	86.7	124.8	45.8	44.3	287.8	5
Co	48.84	51.99	59.13	55.4	58.23	46.19	47.4	44.91	1
Ni	98.82	103.4	64.82	57.14	63.57	35.69	38.28	127.12	2

(continued)

Table 3.9 (continued)

Sample	QX-1		QX-2			QX-3		Standards	Detection limit
	QX-1-1	QX-1-2	QX-2-1	QX-2-2	QX-2-3	QX-3	QX-3	BHVO-2	
Cu	10.68	18.09	83.86	57.16	114.77	55.66	52.20	128.98	1
Zn	68.16	70.56	91.22	92.3	88.71	106.1	114	109.6	2
Ga	11.78	12.4	17.99	17.18	17.2	18.99	18.62	22.32	2
Rb	51.9	48.9	17.7	28.2	24.7	17.5	15.6	9.47	5
Sr	103.9	136.5	146.8	143.4	149.5	171.7	197.4	405	5
Y	11.33	12.02	23.12	22.49	23.77	24.69	25.62	26.99	1
Zr	21.00	24.90	45.20	51.20	43.90	90.00	87.80	173	2
Nb	2.27	2.99	1.69	2.69	1.83	4.72	4.63	18.64	2
Mo	0.19	0.16	0.18	0.43	0.15	0.28	0.23		0.2
Cd	69.57	66.86	147.66	145.62	148.60	144.35	111.06		0.02
In	0.03	0.04	0.06	0.06	0.05	0.06	0.06		0.02
Cs	0.73	0.83	0.24	0.31	0.28	0.46	0.30	0.1	1
Ba	88.60	104.60	64.10	114.50	90.80	222.10	195.20	133	5
La	5.93	4.25	2.91	5.78	3.23	10.35	9.60	15.74	1
Ce	12.16	8.60	7.11	12.29	7.54	21.52	20.42	38.11	1
Pr	1.45	1.11	1.05	1.70	1.13	3.01	2.92	5.36	0.1
Nd	5.52	4.50	5.43	7.55	5.67	12.96	12.81	24.96	0.1
Sm	1.21	1.15	1.90	2.12	1.99	3.24	3.25	6.18	0.1
Eu	0.40	0.40	0.77	0.77	0.88	1.11	1.16	2.1	0.1

(continued)

Table 3.9 (continued)

Sample	QX-1		QX-2			QX-3		Standards	Detection limit
	QX-1-1	QX-1-2	QX-2-1	QX-2-2	QX-2-3	QX-3	QX-3	BHVO-2	
Gd	1.44	1.40	2.61	2.74	2.73	3.68	3.73	6.42	0.1
Tb	0.28	0.28	0.56	0.55	0.58	0.68	0.70	0.98	0.1
Dy	1.83	1.93	3.88	3.72	4.03	4.34	4.50	5.54	0.1
Ho	0.39	0.41	0.82	0.80	0.85	0.88	0.92	1	0.1
Er	1.19	1.25	2.40	2.35	2.47	2.57	2.67	2.64	0.1
Tm	0.19	0.21	0.37	0.37	0.40	0.40	0.40	0.35	0.1
Yb	1.31	1.46	2.59	2.50	2.60	2.58	2.67	2.1	0.1
Lu	0.23	0.25	0.44	0.44	0.43	0.45	0.46	0.29	0.1
Hf	0.87	1.00	1.92	2.33	1.96	3.31	3.01	4.49	0.2
Ta	0.24	0.22	0.12	0.19	0.12	0.32	0.31	1.17	0.1
Pb	2.22	2.36	3.58	3.78	3.29	3.88	3.58	0.04	2
Bi	0.21	0.23	0.18	0.50	0.19	0.06	0.05	1.52	0.05
Th	1.56	0.82	0.91	0.94	0.92	1.40	1.35	1.28	1
U	0.37	0.33	0.31	0.30	0.30	0.28	0.26	0.42	0.2
$^{87}Rb/^{86}Sr$	1.118621	1.033886				0.294147	0.228073		
$^{87}Sr/^{86}Sr$	0.731900	0.731773				0.710044	0.710047		
1δ	0.000007	0.000007				0.000006	0.000009		
$(^{87}Sr/^{86}Sr)i$	0.698040	0.700478				0.701141	0.703143		
$^{147}Sm/^{144}Nd$	0.133005	0.153942				0.151130	0.153238		

(continued)

Table 3.9 (continued)

Sample	QX-1		QX-2			QX-3		Standards	Detection limit
	QX-1-1	QX-1-2	QX-2-1	QX-2-2	QX-2-3	QX-3	QX-3	BHVO-2	
$^{143}Nd/^{144}Nd$	0.511701	0.511979				0.511980	0.512032		
1δ	0.000005	0.000005				0.000005	0.000004		
$(^{143}Nd/^{144}Nd)i$	0.509862	0.509850				0.509890	0.509913		
$\varepsilon_{Nd}(t)$	−1.10	−1.32				−0.54	−0.09		
T_{DM1}	2722	2967				2832	2801		

$Mg^{\#} = 100^{*}$ molar $Mg/(Mg + \sum Fe)$

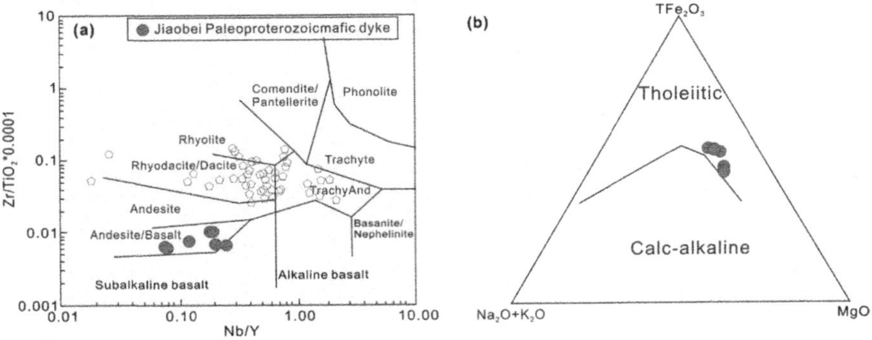

Fig. 3.23 **a** Plot of SiO_2 versus $Zr/TiO_2{*}0.00001$ (after Winchester and Floyd 1976) and **b** AFM diagram (after Irvine and Barager 1971). Jiaobei Paleoproterozoic mafic dykes are from Liang et al. (2020c)

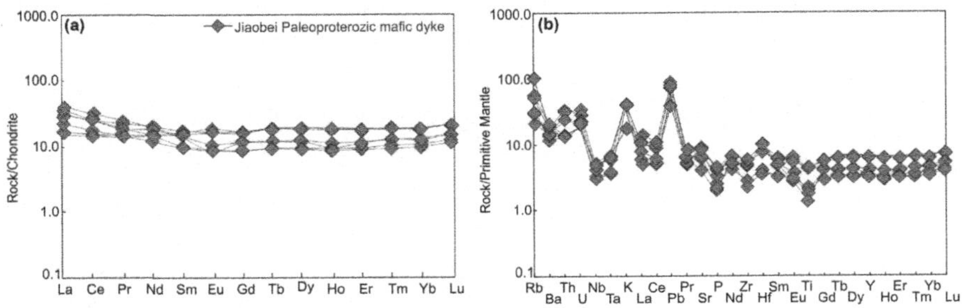

Fig. 3.24 **a** The chondrite-normalized REE contents and **b** primitive mantle-normalized immobile trace element contents diagrams for the Jiaobei mafic dykes. Jiaobei Paleoproterozoic mafic dykes are from Liang et al. (2020c)

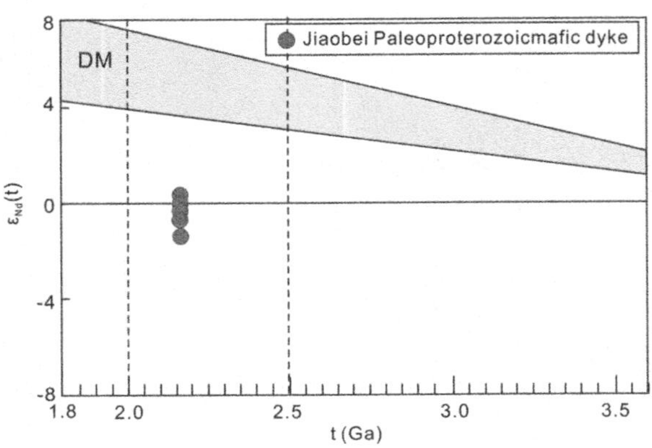

Fig. 3.25 Diagrams of t (Ga) versus εNd(t) for the Jiaobei mafic dykes. Jiaobei Paleoproterozoic mafic dykes are from Liang et al. (2020c)

Ta) and enrichment in LILEs (Fig. 3.24), indicating that the lithospheric mantle of the North China Craton suffered subduction-related factors during this period (Gill 1981; Elliott et al. 1997). As shown in Fig. 3.25, the Nd isotopes of the Jiaobei mafic dyke rocks are different from the lithospheric mantle source rocks of the Archean North China Craton, but are similar to the Paleoproterozoic North China Craton lithospheric mantle source rocks. Wu et al. (2005) have confirmed that the Paleoproterozoic lithospheric mantle in the eastern part of the North China Craton has been altered by subduction or collision-derived crustal materials. The mobility of HFSEs in water-bearing fluids is lower than that of LILEs (Tatsumi 1989; Keppler 1996; Turner et al. 1996), and can generally be used to identify mantle metasomatism. The Jiaobei mafic dyke has a high Rb/Y ratio (~4.58) and low Nb/Y (0.07–0.25), Nb/Zr (0.04–0.12), and Th/Zr (0.02–0.07) ratios, indicating that it originates from the mantle modified by water-bearing fluids (Kepezhinskas et al. 1997). Therefore, the Paleoproterozoic mafic dyke in Jiaobei originated from the lithospheric mantle metasomatised by subduction-related fluids. Many scholars have analysed diamond inclusions in olivine-bearing xenoliths in the Paleozoic Mengyin-Fuxian Kimberlite in the eastern North China Craton to determine the essential characteristics of the lithospheric mantle at the time of their formation (Gao et al. 2002; Wu et al. al. 2006; Zheng et al. 2006; Zhang et al. 2008). These mantle xenoliths show low $\varepsilon_{Nd}(t)$ (-1.7 to -10) and high $^{87}Sr/^{86}Sr$ ratios (0.707–0.715) (Wu et al. 2006; Zhang et al. 2008). The emplacement of the diamond-bearing kimberlite in the Palaeozoic indicates that the lithosphere in eastern North China was generally stable at that time. The large and low-density cratonic roots were characterised by low heat flow and were mainly composed of refractory-rich pyroxene and common pyroxene (Zheng et al. 2002; Zhang et al. 2008).

3.6.2 Lithospheric Mantle Were Enriched by Metasomatism of the Paleo–Pacific Plate Subduction During Mesozoic

Compared with the mantle-derived enclaves in the Paleozoic Kimberlite in the eastern North China Craton, the Shidao and Jiazishan alkaline gabbros (212–210 Ma), which are derived from the Triassic lithospheric mantle in the Sulu orogenic belt, have significantly different geochemical and Sr–Nd isotopic characteristics (Figs. 3.26 and 3.27). This shows that the lithospheric mantle in the eastern block of the North China Craton changed before the Triassic, which may be due to the felsic melt produced by the subduction when the Yangtze Plate subducted below the North China Craton (Zhao et al. 2012). However, the $(^{87}Sr/^{86}Sr)_i$ of the Early Cretaceous mafic dyke in Jiaodong is significantly higher than that of the Late Triassic Sulu dyke (Fig. 3.27). The high water content of the Cretaceous mafic dyke in Jiaodong is not sourced from the subducting Yangtze plate crust, although it is speculated that the lithospheric mantle of the Jiaodong Peninsula has experienced an enrichment process since the Late Triassic.

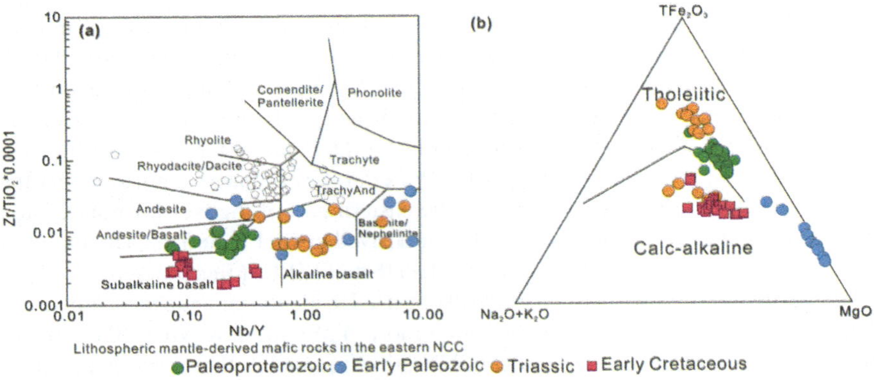

Fig. 3.26 Diagram of Lithospheric mantle-derived mafic rocks for SiO_2-Zr/TiO_2*0.00001 **a** and AFM (**b**). Data source: Paleoproterozoic are from Liang et al. (2020c); Early Paleozoic are from Zhang et al. (2007); Triassic are from Yang et al. (2007); Early Cretaceous are from Deng et al. (2017)

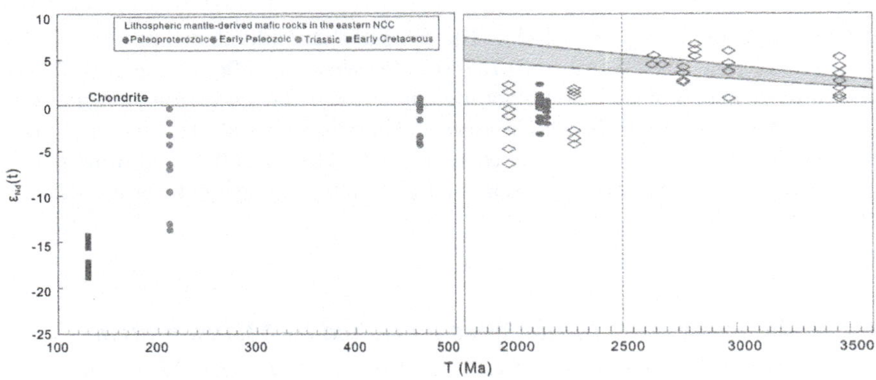

Fig. 3.27 Diagram of Lithospheric mantle-derived mafic rocks for $\varepsilon_{Nd}(t)$-T(Ma). *Data source* Paleoproterozoic are from Liang et al. (2020c); Early Paleozoic are from Zhang et al. (2007); Triassic are from Yang et al. (2007); Early Cretaceous are from Deng et al. (2017)

A study of the whole-rock geochemical characteristics of the Jiaodong mafic dyke revealed that the lithospheric mantle of Jiaodong experienced alterations during the Mesozoic period due to the fluid and melt released by subduction of the Paleo-Pacific plate (Ma et al. 2014, 2016; Deng et al. 2017). The high-Mg clinopyroxene phenocrysts in the Early Cretaceous mafic dykes in Jiaodong crystallised from basaltic magma derived from the metasomatic lithospheric mantle. The characteristics of high-Mg clinopyroxene phenocrysts enriched in LILEs, depleted HFSEs, and enriched in Sr isotopes indicate that the source area of the Jiaodong basic dyke rocks has undergone subduction of seawater-altered oceanic crust and marine sediments. Moreover, the H_2O content of the original magma of the mafic dykes derived from the mantle in the lithosphere in East China also reveals that the mantle source area is metasomatised by subducted oceanic plate containing marine sediment-derived fluids. In addition, the Early Cretaceous Feixian basalt, located in the eastern part of the North China Craton and the western part of the Jiaodong Peninsula also has a higher H_2O content (Xia et al. 2013), which is similar to the coeval Jiaodong mafic dyke. Similar H_2O content and similar geochemical characteristics between the basalt and dykes were observed (Fig. 3.18). The above evidence further shows that the lithospheric mantle of the eastern part of the North China Craton in the Early Cretaceous experienced metasomatism by the source fluids originated from subducting oceanic plates and marine sediments. Since the Late Triassic, the Paleo-Pacific Plate is the only oceanic plate (~178 Ma) that subducted under the eastern part of the North China Craton and eastern China (Li and Li 2007; Sun and Ding 2007; Li et al. 2013). In the Liaodong Peninsula in the north-eastern part of the North China Craton, the Late Jurassic lamprophyre (155 ± 4 Ma) and Early Cretaceous basic rocks (~110 Ma) were confirmed to be derived from partial melting of the enriched mantle influenced by the metasomatism of the Paleo-Pacific Plate (Jiang et al. 2010; Pang et al. 2015). In addition, the Guiyunhua high-magnesium andesite (122 ± 2 Ma) in eastern Liaoning originated from the high-water lithospheric mantle, and its high water content originated from the subduction of the paleo-Pacific plate (Zhao et al. 2007; He 2014; Ma et al. 2016). Therefore, the early Cretaceous Jiaodong lithospheric mantle source area was metasomatised by the source fluids of the subducting paleo-Pacific plate. During the subduction process, the paleo-Pacific plate provided a large amount of water and other volatiles (Liu et al. 2016). Most of these water and volatile components are separated from the subducting plate into the mantle wedge through subduction-related metasomatism and devolatilisation, promoting the melting of the upper mantle wedge (Fischer 2008; Plank et al. 2013). This process increases the water content in the mantle, while imprinting the chemical signature of the mantle rocks by HFSE depletion, LILE enrichment, high $(^{87}Sr/^{86}Sr)i$, and low $\varepsilon Nd(t)$ values.

3.7 Thinning of the Lithosphere Beneath Jiaodong Peninsula as a Result of Continuous Thermo-Mechanical Erosion

During the Mesozoic, the eastern NCC block underwent drastic lithospheric thinning from ~200 km to 80–60 km (Griffin et al. 1998; Xu et al. 2013). A variety of mechanisms for driving the lithospheric thinning have been proposed, including prolonged thermo-mechanical erosion (Xu et al. 2014), rapid lithospheric delamination (Wu et al. 2005), rapid foundering of the lower crust (Gao et al. 2004, 2008; Liu et al. 2009), and a two-stage prolonged thermo-mechanical–chemical erosion (Deng et al. 2017). The model of thermo-mechanical erosion suggests that the lowermost lithosphere was gradually eroded by the upwelling asthenosphere, which represented a slow process lasting at least 100 million years (Menzies et al. 2007). Magmatism related to thermo-mechanical erosion is characterized by the consumption of lithosphere at the beginning of lithospheric thinning. When the lithosphere has thinned to less than ~80 km, partial melting of the asthenosphere is initiated, generating asthenosphere-derived magmas (Xu et al. 2004; Xu 2014). The model of lithospheric delamination, on the other hand, assumes that the thick lithosphere suddenly foundered into the underlying convecting mantle, marking a short event lasting ~10 million years (Menzies et al. 2007). Magmatism related to lithospheric delamination would be rapid and short, and would be characterized by the co-occurrence of lithospheric and asthenospheric mantle-derived magmas, as well as crust- or mantle-derived mixed felsic magmas (Deng et al. 2007; Wu et al. 2008). The model of foundering of lower crust implies that the lower crust foundering resulted in lithospheric extension and asthenospheric upwelling, and this event was initiated by 159 Myr ago and probably was protracted (lasting into the Early Cretaceous) (Gao et al. 2004; Liu et al. 2009). Subsequently, silicic melts originated by melting of the foundered eclogites reacted extensively with the overly mantle peridotite. Primary magma derived from decompressional melting of the hybridized lithospheric mantle (Gao et al. 2008; Liu et al. 2009).

3.7.1 The Variation of Lower Continental Crust Beneath Jiaodong Peninsula

Granites, including I, A, and S-type granites, are produced by the partial melting of continental crust, and have close relationships with continental tectonics, metamorphism, and mineralization (Monecke et al. 2007; Wu et al. 2007; Zeng et al. 2019). Thus, understanding the petrogenesis of these granites can provide information useful for tracing crustal evolution and crust-mantle interaction within the continental lithosphere (Wu et al. 2005; Yang et al. 2007; Zeng et al. 2017). The Mesozoic granitic intrusive rocks are widely distributed in the Jiaodong Peninsula, and can be grouped into Late Jurassic plutons (166–146 Ma), Early Cretaceous intrusions (136–123 Ma), and latest Early Cretaceous intrusions (123–108 Ma) (Goss et al. 2010; Wang et al. 2010a, b; Goldfarb et al. 2014; Tan et al. 2015; Liang et al. 2016; Yang et al. 2016a, b). However, the source and geodynamic background of the late Mesozoic granites in Jiaodong are still poorly understood. Hou et al. (2007) assumed that the late Jurassic Linglong granites found in the northwestern Jiaodong Peninsula were produced by partial melting of Neoarchean lower crust in the NCC. Zhang et al. (2010) determined that the late Jurassic Kunyushan granitoids in the Sulu orogenic belt were originated from partial melting of the continental crust in the post-collisional environment of the subducted Yangtze Block. Jiang et al. (2012) showed that late Jurassic Linglong granites likely originated from partial melting of a thickened lower continental crust with no addition of mantle-derived materials (Ma et al. 2013; Zhao et al. 2018; Li et al. 2019). In contrast, Yang et al. (2003) proposed that the early Cretaceous Guojialing granodiorites were produced by the partial melting of mafic lower crust, which was produced through earlier mantle-derived magmatic underplating. Hou et al. (2007) determined that the Guojialing granodiorites originated from the mixing of delaminated eclogite-rich crust-derived melt and asthenospheric mantle, while Yang et al. (2012) presented a mixed crust-mantle source for the Guojialing granodiorites. Using the geochronology and geochemistry of titanites in the dioritic enclaves and host granodiorites, Jiang et al (2016) determined that the Guojialing intrusion was derived from the mixing of a mafic lower crust-derived dioritic melt and Archean lower crust-derived felsic magma within a shallow reservoir. Zhao et al. (2018) argued that the Guojialing granodiorites resulted from the upward migration of a crustal source from depth, rather than the mixing of mantle-derived materials. Goss et al. (2010) showed that the late stages of granitic batholith emplacement (118–113 Ma) in the Jiaodong Peninsula were produced by the mixing of lower or middle crust-derived felsic magma and mantle-derived materials. Li et al. (2012) assumed that the Gushan granite resulted from multiple magmatic pulses, accompanied by mafic–felsic magma mixing and assimilation. Zhao et al. (2018) argued that the early Cretaceous granitoids (~125–105 Ma) in the Jiaodong Peninsula were produced by partial melting of a shallower crustal source than the earlier two stages of granitic intrusions (e.g. ~170–140 Ma and ~140–125 Ma). Li et al. (2019) assumed that the Aishan complex in the northwestern Jiaodong Peninsula was produced by the partial melting of Neoarchean-Palaeoproterozoic crust mixed with minor mantle-derived materials. Gao et al. (2019) determined that the Laoshan granites resulted from partial melting of granulitic mafic lower crust mixed with underplated basaltic magma (Tang et al. 2014). The obvious variations in the geochemical characteristics of the late Jurassic to latest early Cretaceous felsic intrusive rocks of the Jiaodong Peninsula provide vitally important details for revealing not only the thickening and thinning processes of the lower crust beneath the Jiaodong Peninsula, but also the mechanism that controlled the destruction of the eastern NCC during the late Mesozoic. Although previous studies have discussed the petrogenesis of the three stages of emplaced granites, the sources of these rocks remain poorly constrained and it is unclear whether the lower crust source of the eastern NCC underwent lithospheric thinning. In this book, based on the previously published data of the Late Mesozoic felsic intrusions, we provide synthetically analyses and explanations for the geochemical data from the late Jurassic to early Cretaceous felsic intrusive rocks on the Jiaodong Peninsula. The geochemical variations described herein constrain the change in crustal source for the late Mesozoic granites and thinning of the lower crust, which further constrain the mechanism of the destruction of eastern NCC, including Jiaodong Peninsula.

This chapter are based on a comprehensive compilation of previously published geochronological and geochemical data obtained from the felsic intrusive rocks on the Jiaodong Peninsula; including Late Jurassic granites; 166–144 Ma) are from Zhang et al. 2010; Jiang et al. 2013; Ma et al. 2013; Zhao et al. 2016, 2018 and Li et al. 2019; Early Cretaceous granodiorites (136–123 Ma) are from Zhang et al. 2010; Yang et al. 2012; Wang et al. 2014; Zhao et al. 2018 and Li et al. 2019; latest Early Cretaceous granites (123–108 Ma) are from Huang et al. 2006; Goss et al. 2010; Zhang et al. 2010; Tang et al. 2014; Li et al. 2012, 2014; Yan and Shi 2016; Zhao et al. 2018; Gao

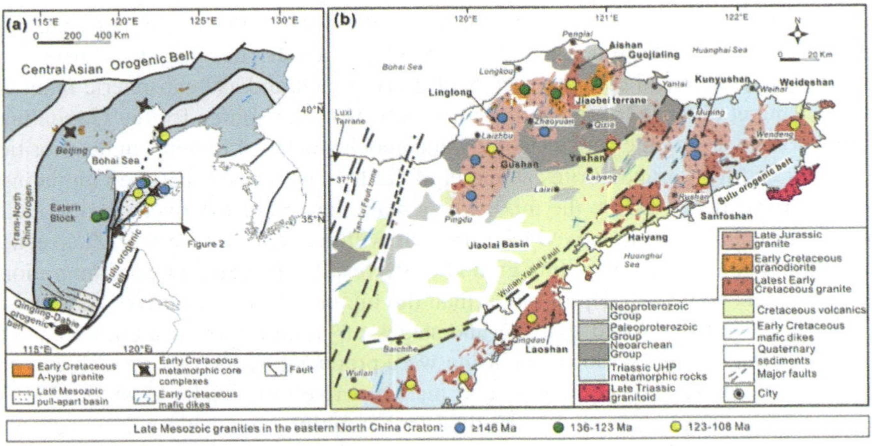

Fig. 3.28 **a** Geologic map of the eastern North China Craton showing locations of Late Mesozoic felsic rocks, Early Cretaceous A-type granites, metamorphic core complexes, pull-apart basins, and mafic dykes (modified after Yang et al. 2007). **b** Simplified geologic map of the Jiaodong Peninsula (modified after Deng et al. 2017), showing the distribution of Late Mesozoic felsic rocks on the Jiaodong Peninsula

et al. 2019; Li et al. 2019 and Yan et al. 2019 (Fig. 3.28). These datasets reveal major chemical variations over time, thus constituting a tool for tracking possible changes in the magmatic sources and lithospheric destruction. Zircon U–Pb ages, Hf–O isotopic data, whole-rock major and trace elemental concentrations, and Sr–Nd–Pb isotopic data were collected from previous studies and are listed in the Tables 3.10 and 3.11 and showed in Figs. 3.29, 3.30, 3.31, 3.32, 3.33 and 3.34.

3.7.1.1 The Variation of Crust Source for the Mesozoic Felsic Intrusions

(1) The Late Jurassic intrusive rocks (166–144)

The Late Jurassic intrusive rocks mainly include medium-grained (i.e. Linglong granite) and coarse-grained (i.e. Luanjiahe granite) biotite granites in the Jiaobei terrane (166–146 Ma; Zhang et al. 2010; Jiang et al. 2012; Ma et al. 2013; Yang et al. 2014; Li et al. 2019) (Fig. 3.35), and biotite-potassium feldspar granites (i.e. Kunyushan granite) in the northern Sulu orogenic belt (157–146 Ma, Zhang et al. 2010; Zhao et al. 2016). Monzogranite (LLW50-7) from the Linglong gold deposit in the Jiaobei terrane was sampled for zircon U–Pb dating. Fourteen zircons showed narrow ranges of Th (50–342 ppm) and U (81–1460 ppm) contents, with Th/U ratios of 0.17–1.07, and yielded a weighted mean age of 144.8 ± 2.7 (2σ, MSWD = 1.7) (Fig. 3.36).

The Late Jurassic felsic intrusive rocks (166–144 Ma) all exhibit high SiO_2 and K_2O contents, mostly plotting in the high-K calc-alkaline series, with minor

Table 3.10 The summarize of the Late Mesozoic felsic rocks in the Jiaodong gold province (after Liang et al. 2020a)

Group	Late jurassic	Early cretaceous	Latest early cretaceous
Age (Ma)	166–146 Ma	136–123 Ma	123–108 Ma
Rock type	Medium-grained (i.e. Linglong granite) and coarse-grained (i.e. Luanjiahe granite) biotite granites in the Jiaobei terrane; and biotite-potassium feldspar granites (i.e. Kunyushan granite) in the northern Sulu orogenic belt	Medium- to coarse-grained granodiorites (i.e. Guojialing granodiorites)	Aishan complex (e.g. granodiorites, syenogranites, quartz syenogranites, alkali granites, and monzodiorites), the Gushan granite, and the Yashan coarse-grained monzogranite in the Jiaobei terrane; the Weideshan granite, the Sanfoshan complex (e.g. monzonites, quartz monzonites, granodiorites, and granites), the Haiyang K-feldspar granite, the Laoshan granitoids and syenogranite, and the Wulian-Dazhuangshan granite in the Sulu Orogenic Belt
Major elements (wt%)	SiO_2 = 68.8–75. K_2O = 2.84–5.39, Al_2O_3 = 12.9–16.9, Na_2O = 2.79–5.12, $Mg^{\#}$ = 11.0–48.8, MgO = 0.05–1.37, TiO_2 = 0.03–0.29, CaO = 0.33–2.93	SiO_2 = 62.4–77.7, K_2O = 1.33–5.39, Al_2O_3 = 11.9–17.5, Na_2O = 3.4–5.92 $Mg^{\#}$ = 22.9–71.4, MgO = 0.14–2.13, TiO_2 = 0.1–0.72, CaO = 0.43–3.73	SiO_2 = 56.1–77.6, K_2O = 0.04–7.75, Al_2O_3 = 9.22–17.4, Na_2O = 1.85–5.07, $Mg^{\#}$ = 5.05–64.6, MgO = 0.07–3.87, TiO_2 = 0.07–0.81, CaO = 0.43–3.73
Type	High-K calc-alkaline series with minor shoshonitic series	High-K calc-alkaline series with minor calc-alkaline	High-K calc-alkaline to shoshonitic series
Type	Peraluminous series	Metaluminous and slightly peraluminous	Metaluminous to peraluminous
Trace elements	Sr/Y = 20.4–308 (average 84.5), Dy/Yb = 1.03–3.36 (average 1.94), La/Yb = 7.48–104 (average 44.8)	Sr/Y = 62.6–378 (average 168), Dy/Yb = 1.45–4.41 (average 2.98), La/Yb = 8.46–206 (average 96.3)	Sr/Y = 0.27–108 (average 25.3), Dy/Yb = 0.39–2.49 (average 1.54), La/Yb = 2.59–176 (average 38.7)
$(^{87}Sr/^{86}Sr)i$	0.707636–0.713900	0.710175–0.712539	0.705400–0.724600
$\varepsilon_{Nd}(t)$	−22.1 to −10.9	−22.3 to −11.5	−22.3 to −3.0
$(^{206}Pb/^{204}Pb)i$	16.550–17.060	16.880–17.330	16.119–17.342

(continued)

Table 3.10 (continued)

Group	Late jurassic	Early cretaceous	Latest early cretaceous
$(^{207}Pb/^{204}Pb)i$	15.380–15.470	15.440–15.544	15.386–15.505
$(^{208}Pb/^{204}Pb)i$	36.920–37.780	37.590–38.282	36.194–37.930
$\varepsilon_{Hf}(t)$	−43.5 to −11.3	−25.1 to −11.6	−25.5 to −5.4
TDM2	2843–1914 Ma	3221–1917 Ma	3900–1515 Ma
References	Zhang et al. (2010), Jiang et al. (2013), Ma et al. (2013), Zhao et al. (2016, 2018), Li et al. (2019)	Zhang et al. (2010), Yang et al. (2012), Wang et al. (2014), Zhao et al.(2018), Li et al. (2019)	Huang et al. (2006), Goss et al. (2010), Zhang et al. (2010), Tang et al. (2014), Li et al. (2012, 2014), Yan and Shi (2016), Zhao et al. (2018), Gao et al. (2019), Li et al. (2019), Yan et al. (2019)

shoshonitic series rocks. They also have relatively high Al_2O_3 and Na_2O contents, with high A/CNK [molar ratio $Al_2O_3/(Na_2O + K_2O)$] values, which places them in the peraluminous series. They have low $Mg^{\#}$ values (11.0–48.8) and limited MgO contents (0.05–1.37 wt%). They have high Sr (145–1700 ppm, average 574 ppm) and low Y (1.20–14.8 ppm, average 6.87 ppm) and Yb (0.25–3.13 ppm, average 0.73 ppm) compositions, with high Sr/Y (20.4–308, average 84.5), Dy/Yb (1.03–3.36, average 1.94), and La/Yb (7.48–104, average 44.8) ratios, showing adakitic features. In addition, these granites are depleted in high field strength elements (HFSE, such as Nb, Ta, Zr, and Hf) and heavy rare earth elements (HREEs), and are enriched in large ion lithophile elements (LILEs) and light rare earth elements (LREEs), along with lack of a Eu anomaly ($\delta Eu = 0.13–0.89$). They have high $(^{87}Sr/^{86}Sr)_i$ (0.707636–0.713900) and low $\varepsilon_{Nd}(t)$ (−22.1 to −10.9) values and narrow ranges of $(^{206}Pb/^{204}Pb)_i$ (16.550–17.060), $(^{207}Pb/^{204}Pb)_i$ (15.380–15.470), and $(^{208}Pb/^{204}Pb)_i$ (36.920–37.780) isotopic ratios. Zircons from the Late Jurassic granites exhibit $\varepsilon_{Hf}(t)$ values of −43.5 to −11.3, with two-stage model ages (T_{DM2}) of 2843–1914 Ma (Figs. 3.29, 3.30, 3.31, 3.32, 3.33 and 3.34).

Late Jurassic granites in the Jiaodong Peninsula have high Sr/Y ratios and La/Yb ratios, and most of the samples plot in the adakite region (Fig. 3. 37). Several genetic models for the generation of the adakitic features have been provided: (1) partial melting of a subducting oceanic slab (Defant and Drummond 1990; Martin 1999); (2) partial melting of delaminated lower crust (Xu et al. 2002; Gao et al. 2008); (3) crustal assimilation and/or fractional crystallization (AFC) from basaltic magmas (Castillo et al. 1999; Castillo 2012). (4) deep fractionation processes (involving garnet or amphibole) of hydrous basaltic melts (Richards and Kerrich 2007; Chiaradia et al. 2009) and (5) partial melting of thickened lower crust (Chung et al. 2003; Ma et al. 2013; Long et al. 2015) (Tables 3.12 and 3.13).

Firstly, the typical adakites is recognized to be originated from partial melting of subducting basaltic oceanic crust (Defant and Drummond 1990). However, these

Table 3.11 LA-ICP-MS and/or SHRIMP zircon U–Pb ages of the Late Mesozoic felsic rocks in the Jiaodong Peninsula. (after Liang et al. 2020a)

Location	Rock stype	Method	Age (Ma)	References
Late Jurassic granites				
Kunyushan	Granite	LA-ICP-MS U–Pb	146 ± 4	Zhang et al. (2010)
Kunyushan	Granite	LA-ICP-MS U–Pb	149 ± 5	Zhang et al. (2010)
Kunyushan	Granite	LA-ICP-MS U–Pb	146 ± 4	Zhang et al. (2010)
daliuhang	Granite	LA-ICP-MS U–Pb	157 ± 2	Zhang et al. (2010)
linglong	Granite	LA-ICP-MS U–Pb	147 ± 3	Zhang et al. (2010)
Jiaobei terrain	Biotite granite	LA-ICP-MS U–Pb	150 ± 2	Jiang et al. (2013)
Jiaobei terrain	Biotite granite	LA-ICP-MS U–Pb	156 ± 1	Jiang et al. (2013)
Jiaobei terrain	Biotite granite	LA-ICP-MS U–Pb	166 ± 4	Jiang et al. (2013)
Jiaobei terrain	Biotite granite	LA-ICP-MS U–Pb	166 ± 5	Jiang et al. (2013)
Jiaobei terrain	Biotite granite	LA-ICP-MS U–Pb	149 ± 2	Jiang et al. (2013)
Jiaobei terrain	Granite	LA-ICP-MS U–Pb	158 ± 2	Yang et al. (2012)
Jiaobei terrain	Granite	LA-ICP-MS U–Pb	157 ± 2	Yang et al. (2012)
Jiaobei terrain	Granite	LA-ICP-MS U–Pb	159 ± 2	Yang et al. (2012)
Jiaobei terrain	Granite	LA-ICP-MS U–Pb	159 ± 1	Yang et al. (2012)
Jiaojia gold deposit	Granite	LA-ICP-MS U–Pb	160 ± 2.9	Ma et al. (2013)
Jiaojia gold deposit	Granite	LA-ICP-MS U–Pb	157 ± 3	Ma et al. (2013)
Sulu orogenic belt	Granite	LA-ICP-MS U–Pb	156.6 ± 0.6	Zhao et al. (2016)
Sulu orogenic belt	Granite	LA-ICP-MS U–Pb	146.5 ± 0.5	Zhao et al. (2016)
Sulu orogenic belt	Granite	LA-ICP-MS U–Pb	151.3 ± 1.3	Zhao et al. (2016)
Sulu orogenic belt	Granite	LA-ICP-MS U–Pb	155 ± 1.7	Zhao et al. (2016)

(continued)

Table 3.11 (continued)

Location	Rock stype	Method	Age (Ma)	References
Jiaobei terrain	Granite	LA-ICP-MS U–Pb	144.9 ± 0.8	Zhao et al. (2016)
Jiaobei terrain	Granite	LA-ICP-MS U–Pb	158 ± 2	Li et al. (2019)
Jiaobei terrain	Granite	LA-ICP-MS U–Pb	166 ± 1	Li et al. (2019)
Jiaobei terrain	Granite	LA-ICP-MS U–Pb	161.4 ± 1.7	Zhao et al. (2018)
Early Cretaceous granodiorites				
Guojialing	Granodiorite	LA-ICP-MS U–Pb	121 ± 2	Zhang et al. (2010)
Guojialing	Granodiorite	LA-ICP-MS U–Pb	133 ± 3	Zhang et al. (2010)
Guojialing	Granodiorite	LA-ICP-MS U–Pb	129 ± 1	Yang et al. (2012
Guojialing	Granodiorite	LA-ICP-MS U–Pb	129 ± 2	Yang et al. (2012)
Guojialing	Granodiorite	LA-ICP-MS U–Pb	130 ± 2	Li et al. (2019)
Guojialing	Granodiorite	LA-ICP-MS U–Pb	128 ± 1	Li et al. (2019)
Sanshandao gold deposit	Granodiorite	LA-ICP-MS U–Pb	130 ± 1.2	Zhao et al. (2018)
Xinli gold deposit	Granodiorite	LA-ICP-MS U–Pb	130.2 ± 0.4	Zhao et al. (2018)
Xinchen gold deposit	Granodiorite	LA-ICP-MS U–Pb	128 ± 1	Wang et al. (2014)
Xinchen gold deposit	Granodiorite	LA-ICP-MS U–Pb	128 ± 1	Wang et al. (2014)
Xinchen gold deposit	Granodiorite	LA-ICP-MS U–Pb	132 ± 1	Wang et al. (2014)
Xinchen gold deposit	Granodiorite	LA-ICP-MS U–Pb	126 ± 1	Wang et al. (2014)
Latest Early Cretaceous granites				
Aishan	Monzogranite	SHRIMP U-Pb	116 ± 2	Goss et al. (2010)
Aisha	Monzogranite	SHRIMP U-Pb	125 ± 3	Goss et al. (2010)
Gushan	Granite	LA-ICP-MS U–Pb	118.7 ± 0.9	Li et al. (2012)

(continued)

Table 3.11 (continued)

Location	Rock stype	Method	Age (Ma)	References
Gushan	Granite	LA-ICP-MS U–Pb	120 ± 1.1	Li et al. (2012)
Gushan	Granite	LA-ICP-MS U–Pb	118.5 ± 1	Li et al. (2012)
Yashan	Monzonite	SHRIMP U-Pb	116 ± 1	Goss et al. (2010)
Yashan	Monzogranite	SHRIMP U-Pb	113 ± 2	Goss et al. (2010)
Aishan	Granodiorite	SIMS U-Pb	118 ± 1	Li et al. (2019)
Aishan	Granodiorite	SIMS U-Pb	118 ± 1	Li et al. (2019)
Aishan	Granodiorite	SIMS U-Pb	115 ± 1	Li et al. (2019)
Aishan	Granodiorite	LA-ICP-MS U-Pb	117 ± 1	Li et al. (2019)
Sanfoshan	Monzonite	LA-ICP-MS U-Pb	114 ± 3	Zhang et al. (2010)
Sanfoshan	Granite	LA-ICP-MS U-Pb	116 ± 3	Zhang et al. (2010)
Sanfoshan	Monzogranite	SHRIMP U-Pb	118 ± 1	Goss et al. (2010)
Sanfoshan	Syenogranite	SHRIMP U-Pb	117 ± 1	Goss et al. (2010)
Sanfoshan	K-feldspar granite	LA-ICP-MS U-Pb	118.1 ± 2.8	Tang et al. (2014)
Wulianshan	Granite	LA-ICP-MS U-Pb	118 ± 2	Huang et al. (2006)
Wulianshan	Granite	SHRIMP U-Pb	121 ± 3	Huang et al. (2006)
Wulianshan	Granite	LA-ICP-MS U-Pb	116 ± 2	Huang et al. (2006)
Wulianshan	Granite	LA-ICP-MS U-Pb	119.1 ± 0.9	Yan et al. (2019)
Wulianshan	Granite	LA-ICP-MS U-Pb	122.3 ± 0.58	Yan et al. (2019)
Xiozhushan	Granite	LA-ICP-MS U-Pb	114.2 ± 0.8	Yan et al. (2019)
Dazhuangshan	Granite	LA-ICP-MS U-Pb	108.9 ± 1.9	Yan et al. (2019)

(continued)

Table 3.11 (continued)

Location	Rock stype	Method	Age (Ma)	References
Haiyang	K-feldspar granite	LA-ICP-MS U-Pb	116.8 ± 1.7	Li et al. (2014)
Haiyang	K-feldspar granite	LA-ICP-MS U-Pb	115.8 ± 2.2	Li et al. (2014)
Laoshan	Granodiorite	LA-ICP-MS U-Pb	123.4 ± 1.6	Yan and Shi (2016)
Laoshan	Granodiorite	LA-ICP-MS U-Pb	115.7 ± 1.4	Yan and Shi (2016)
Laoshan	Quartz monozite	LA-ICP-MS U-Pb	116.1 ± 1.2	Yan and Shi (2016)
Laoshan	Quartz monozite	LA-ICP-MS U-Pb	115.9 ± 1.2	Yan and Shi (2016)
Laoshan	Alkali-feldspar granite	LA-ICP-MS U-Pb	120.4 ± 1.6	Gao et al. (2019)
Laoshan	Monzonitic granite	LA-ICP-MS U-Pb	119.4 ± 1.3	Gao et al. (2019)
Aishan	Granite	LA-ICP-MS U-Pb	107.9 ± 0.5	Zhao et al. (2018)
Yashan	Granite	LA-ICP-MS U-Pb	114.4 ± 0.3	Zhao et al. (2018)
Hiayang	Granite	LA-ICP-MS U-Pb	107.2 ± 0.5	Zhao et al. (2018)
Weideshan	Granite	LA-ICP-MS U-Pb	114 ± 0.1	Zhao et al. (2018)

Late Jurassic granites in the Jiaodong Peninsula have much lower Nd and higher Sr isotopic compositions than those of typical adakites (Fig. 3.32a), and show different Pb isotope (Fig. 3.32b–c) and trace element compositions with MORB (Fig. 3.31b). Thus, the model of partial melting of a subducting the oceanic crust could be excluded. Secondly, adakitic features also could be derived from partial melting of delaminated lower crust (Xu et al. 2002; Gao et al. 2004). In this model, lower crustal rocks delaminated into the mantle, and followed by interaction with mantle peridotites, generating high-Mg adakitic rocks that have higher MgO (>3 wt%) and $\varepsilon_{Nd}(t)$ (−10 to +1), and lower $(^{87}Sr/^{86}Sr)_i$ (0.7051–0.7075) than Late Jurassic granites. Thus, Late Jurassic granites also didn't derive from delaminated lower crust source. Thirdly, the lack of correlation between $(^{87}Sr/^{86}Sr)_i$ and SiO_2 contents of Late Jurassic granites and small isotopic variety of Sr and Nd isotopes could exclude crustal contamination. Besides, AFC processes would cause the magmas toward low Sr/Y and negative Eu anomalies (Ma et al. 2015), which is inconsistent with these granites. Fourthly, deep and high-pressure fractionation process of garnet could generate the positive correlation between Sr/Y and Dy/Yb with SiO_2 and overall low HREE contents (Chiaradia et al. 2009), which also is insistent with Late Jurassic granites. Moreover,

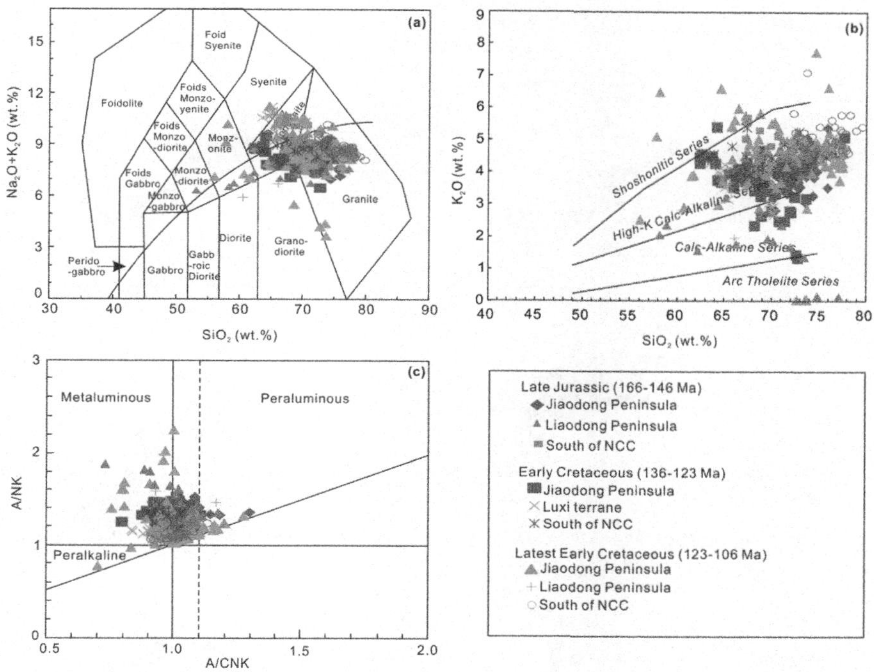

Fig. 3.29 Plots of **a** $Na_2O + K_2O$ versus SiO_2, **b** K_2O versus SiO_2 (after Le et al. 1989), and **c** A/NK [molar ratio $Al_2O_3/(Na_2O + K_2O)$] versus A/CNK [molar ratio $Al_2O_3/(CaO + Na_2O + K_2O)$] (Maniar and Piccoli 1989) for Late Mesozoic felsic rocks on the Jiaodong Peninsula and eastern NCC. *Data source* are listed Table 3.1

the contemporary basaltic magma couldn't be found in the Jiaodong Peninsula. So, the model of deep fractionation processes of hydrous basaltic melts also didn't explain the source of Late Jurassic granites.

Alternatively, their geochemical characteristics are similar to C-type adakitic rocks that are derived from thickened lower continental crust (LCC) (Li et al. 2013). According to plots of MgO versus SiO_2 and TiO_2 versus SiO_2, the late Jurassic granites plot in the ranges for thickened LCC (Fig. 3.30a and b), which resemble the late Mesozoic thickened LCC-derived adakitic plutons in the eastern NCC (Xu et al. 2012; Yang et al. 2016). Late Jurassic granites have high SiO_2 (68.8–75.9 wt%) and low MgO (0.14–2.13 wt%) contents, high initial $^{87}Sr/^{86}Sr$ ratios (0.708–0.714), and negative $\varepsilon_{Nd}(t)$ (−22.1 to −10.9) and zircon $\varepsilon_{Hf}(t)$ (−43.5 to −11.3) values, indicating a continental crust-derived or highly evolved magma. The two-stage Nd model ages (T_{DM2}) of these granites range from 2732 to 2047 Ma (Fig. 3.34a). Magmatic zircons from late Jurassic granites exhibit two-stage Hf model ages (T_{DM2}) ranging from 3221–1917 Ma (Fig. 3.34b). Neoarchean and Palaeoproterozoic inherited zircons (2848–1615 Ma) were also observed in these granites. The measured ages are consistent with granulite xenoliths captured in late Cretaceous mafic dykes

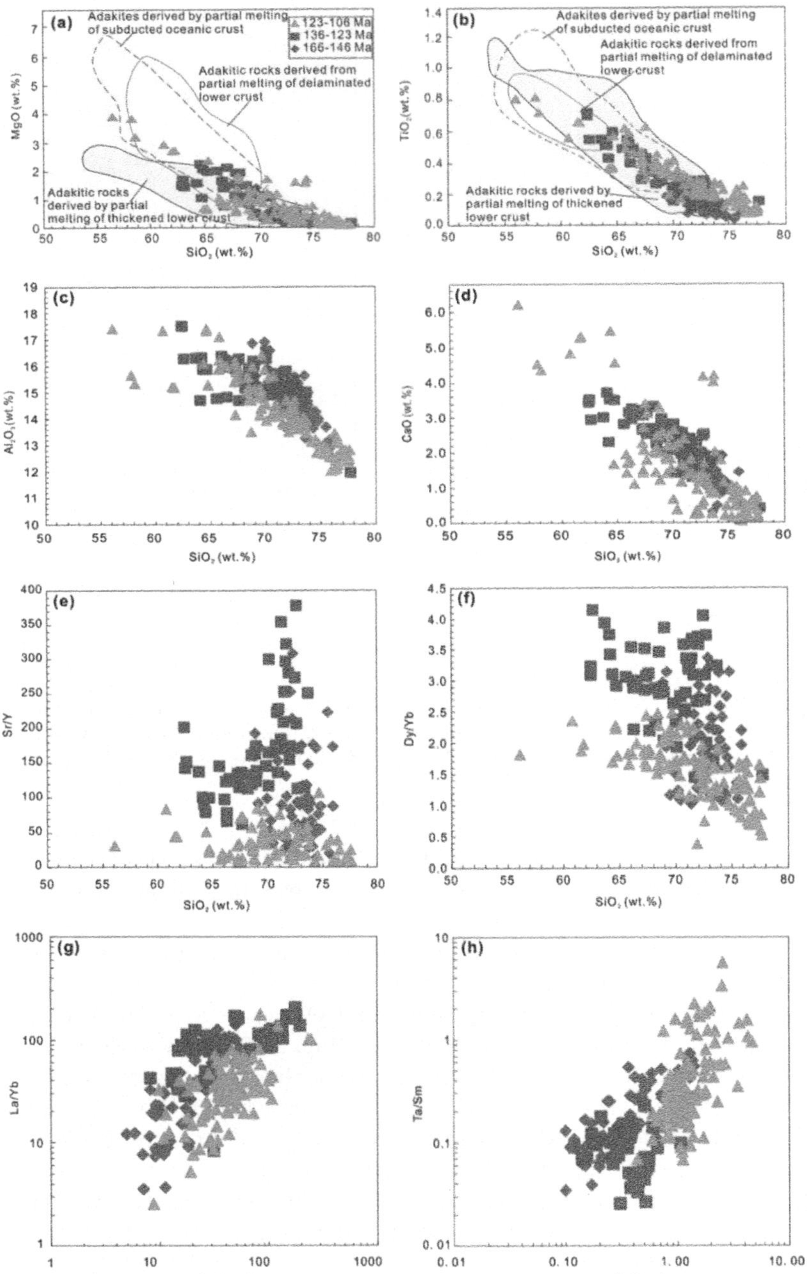

Fig. 3.30 **a–d** Harker diagrams of selected major and trace element contents, **e** La versus La/Yb, and **f** Ta versus Ta/Sm for Late Mesozoic felsic rocks on the Jiaodong Peninsula. Data source are listed Table 3.1

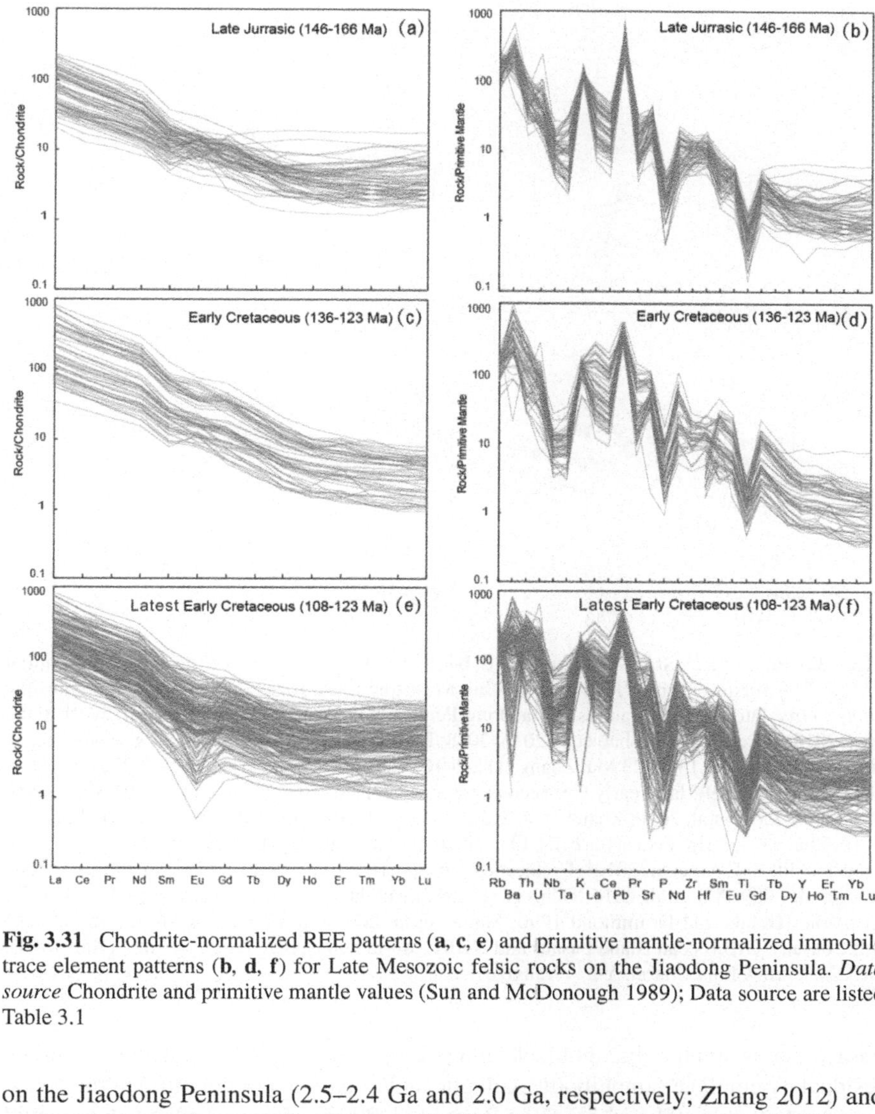

Fig. 3.31 Chondrite-normalized REE patterns (**a**, **c**, **e**) and primitive mantle-normalized immobile trace element patterns (**b**, **d**, **f**) for Late Mesozoic felsic rocks on the Jiaodong Peninsula. *Data source* Chondrite and primitive mantle values (Sun and McDonough 1989); Data source are listed Table 3.1

on the Jiaodong Peninsula (2.5–2.4 Ga and 2.0 Ga, respectively; Zhang 2012) and ancient basement beneath the Jiaodong Peninsula (2.9 to 2.5 Ga; 2.2 to 1.9 Ga and ~1.9 Ga; Zhai and Bian 2000; Wan et al. 2006). Moreover, Li et al. (2019) showed that the Linglong granites have high $\delta^{18}O$ values (5.72–8.34‰), indicating an Archean lower crustal source. Thus, the late Jurassic granites originated from partial melting of the thickened Archean lower crust of the eastern NCC.

Given at high-pressure (>~1.0 GPa) HREE (e.g. Y and Yb) will preferentially partitioned into garnet or amphibole, but Sr and La will enter the liquid phase, generating high Sr/Y and La/Yb ratios (Profeta et al. 2015; Zhu et al. 2017b), Late Jurassic granites show adakitic characteristics, including high Sr/Y and La/Yb ratios, and low HREE compositions, which suggest a magma source that contains residual

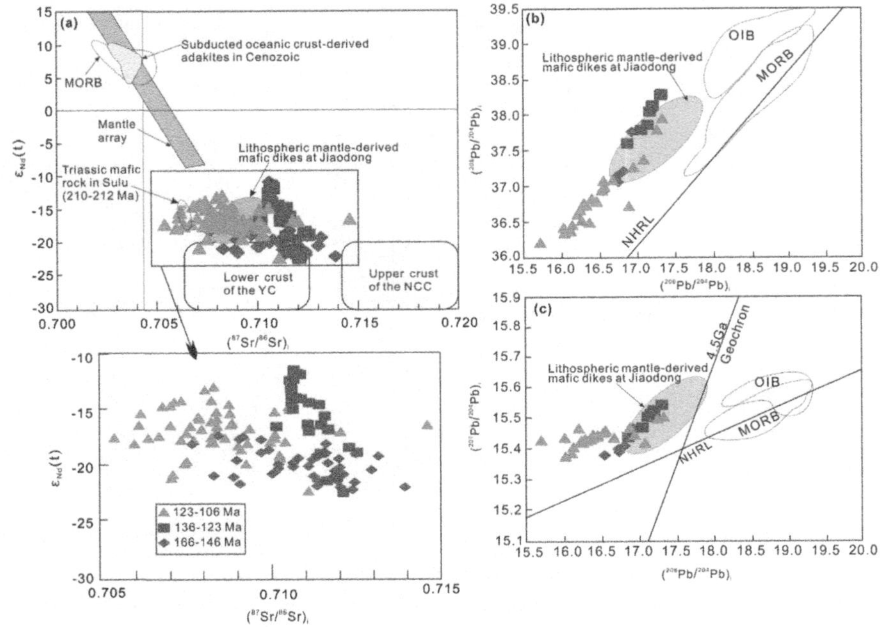

Fig. 3.32 Plots of **a** $(^{87}Sr/^{86}Sr)_i$ versus $\varepsilon_{Nd}(t)$, **b** initial $^{206}Pb/^{204}Pb$ versus $^{208}Pb/^{204}Pb$, and **c** initial $^{206}Pb/^{204}Pb$ versus $^{207}Pb/^{204}Pb$ values of Late Mesozoic felsic rocks on the Jiaodong Peninsula. *Data source* late Jurassic granites in Jiaodong Peninsula (166–146 Ma; Zhang et al. 2010; Jiang et al. 2013; Ma et al. 2013; Zhao et al. 2016, 2018; Li et al. 2019); early Cretaceous granodiorites in Jiaodong Peninsula (136–123 Ma; Zhang et al. 2010; Yang et al. 2012; Wang et al. 2014; Zhao et al. 2018; Li et al. 2019); later early Cretaceous granites in Jiaodong Peninsula (123–108 Ma: Huang et al. 2006; Goss et al. 2010; Zhang et al. 2010; Tang et al. 2014; Li et al. 2012, 2014; Yan and Shi 2016; Zhao et al. 2018; Gao et al. 2019; Li et al. 2019; Yan et al. 2019); Triassic mafic rocks in Sulu orogenic dikes (Zhao et al. 2012); Cretaceous lithospheric mantle-derived mafic dikes at Jiaodong (Liu et al. 2009; Ma et al. 2014; Deng et al. 2017); subducted oceanic crust-derived adakites in Cenozoic (Defant and Drummond 1990; Sajona et al. 2000); and fields for MORB (mid-ocean ridge basalt), OIB (ocean island basalt) and NHRL (north hemisphere reference line) (Hart 1984). NCC = North China Craton; YC = Yangtze

garnet and/or amphibolite, but lacks plagioclase. Garnet could be stable within the residual assemblage (amphibolite–eclogite and/or garnet–amphibolite) in thickened continent crust (>40 km; ~1.2 GPa; Rapp et al. 2003). However, plagioclase would disappear at pressures above 1.2–1.5 GPa (Rapp et al. 2003). From these combined data, we postulate that the late Jurassic granites may have been sourced from a plagioclase-free amphibolite–eclogite or a garnet–amphibolite crustal source in the eastern NCC, with an overall thickness of >40 km.

(2) **The Early Cretaceous granodiorites (136–123 Ma)**

The Early Cretaceous felsic intrusive rocks (136–123 Ma), located in the north-eastern Jiaobei terrane, are composed of medium- to coarse-grained granodiorites

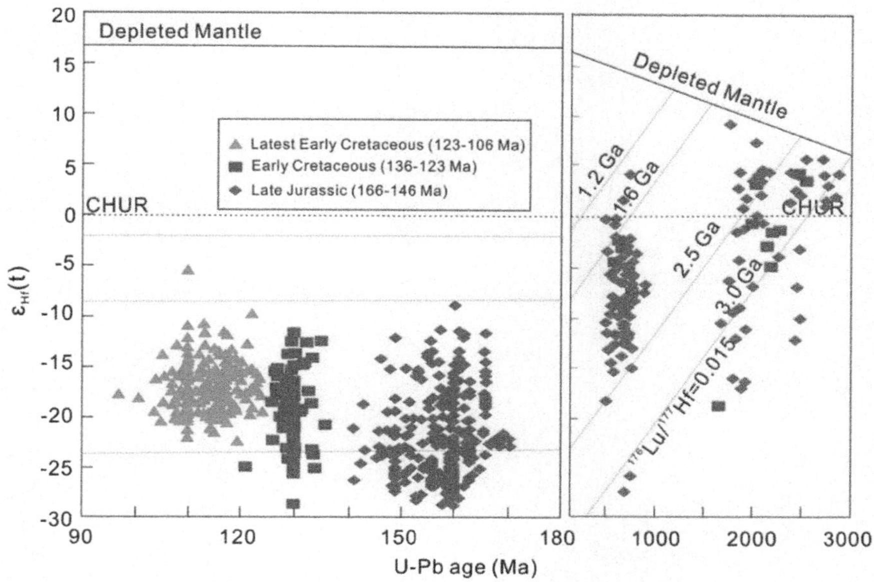

Fig. 3.33 Plots of zircon $\varepsilon_{Hf}(t)$ values versus ages of Late Mesozoic felsic rocks on the Jiaodong Peninsula. *Data source* late Jurassic granites (166–146 Ma; Jiaodong Peninsula- Zhang et al. 2010; Jiang et al. 2013; Ma et al. 2013; Zhao et al. 2016, 2018; Li et al. 2019), early Cretaceous granodiorites (136–123 Ma; Jiaodong Peninsula- Zhang et al. 2010; Yang et al. 2012; Wang et al. 2014; Zhao et al. 2018; Li et al. 2019), and later early Cretaceous granites (123–108 Ma: Jiaodong Peninsula-Huang et al. 2006; Goss et al. 2010; Zhang et al. 2010; Tang et al. 2014; Li et al. 2012, 2014; Yan and Shi 2016; Zhao et al. 2018; Gao et al. 2019; Li et al. 2019; Yan et al. 2019). CHUR = Chondrite uniform reservoir

(i.e. Guojialing granodiorites). They were emplaced between 136–123 Ma (zircon U-Pb age; Zhang et al. 2010; Yang et al. 2012; Wang et al. 2014; Zhao et al. 2018; Li et al. 2019), and also have high SiO_2, Na_2O, and K_2O contents, mostly plotting in the high-K calc-alkaline series with minor calc-alkaline series rocks. Most of these granitoids are metaluminous and slightly peraluminous. They have large ranges of $Mg^{\#}$ (22.9–71.4) and limited MgO contents (0.14–2.13 wt%). This stage of intrusive rocks also shows adakitic features with higher Sr (42.9–2302 ppm, average 1159 ppm) contents and lower Y (2.41–39.0 ppm, average 8.92 ppm) and Yb (0.16–3.71 ppm, average 0.67 ppm) contents, along with higher Sr/Y (62.6–378, average 168), Dy/Yb (1.45–4.41, average 2.98), and La/Yb (8.46–206, average 96.3) ratios than the Late Jurassic intrusive rocks. These granites exhibit HFSE and HREE depletions and LILE and LREE enrichments, along with the lack of a Eu anomaly ($\delta Eu = 0.06$–0.64). They have high $(^{87}Sr/^{86}Sr)_i$ (0.710175–0.712539) and low $\varepsilon_{Nd}(t)$ (−22.3 to −11.5) values, with narrow ranges of $(^{206}Pb/^{204}Pb)_i$ (16.880–17.330), $(^{207}Pb/^{204}Pb)_i$ (15.440–15.544), and $(^{208}Pb/^{204}Pb)_i$ (37.590–38.282) isotopic ratios. Zircons from Early Cretaceous granites exhibit $\varepsilon_{Hf}(t)$ values of −25.1 to −11.6, with T_{DM2} ages of 3221–1917 Ma (Figs. 3.29, 3.30, 3.31, 3.32, 3.33 and 3.34).

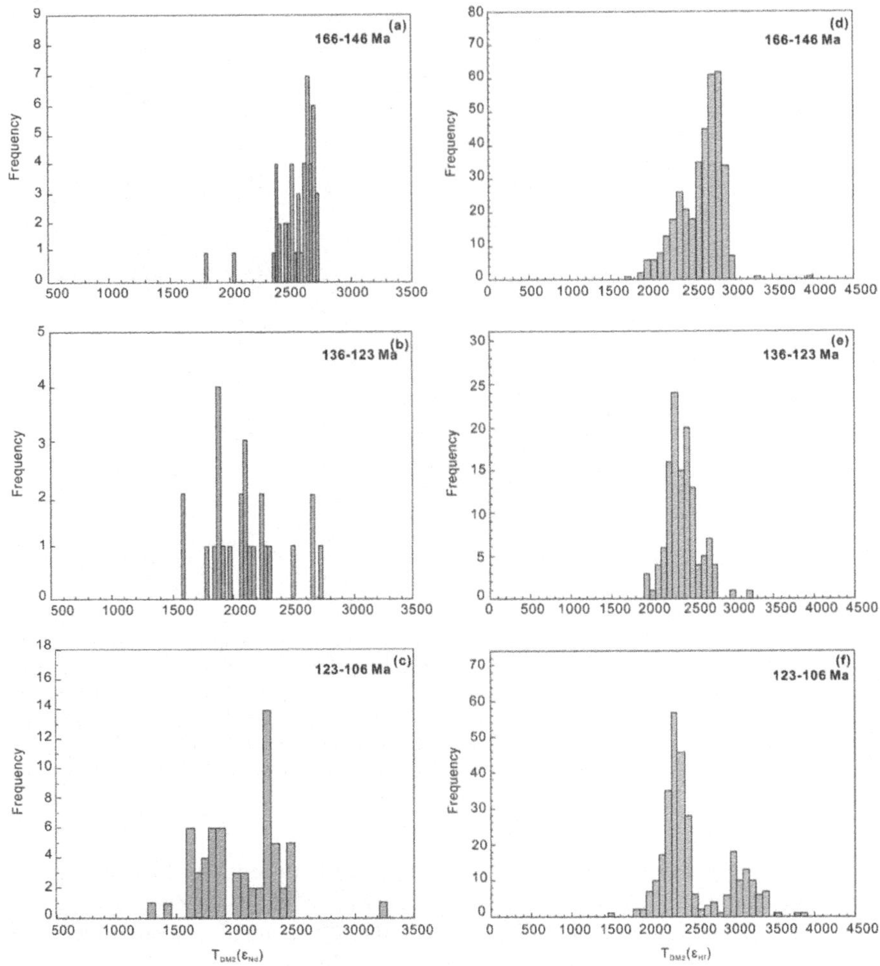

Fig. 3.34 Zircon two-stage Nd model ages (TDM2) **a–c** and two-stage Hf model ages (TDM2) **d–f** for Late Mesozoic felsic rocks on the Jiaodong Peninsula. *Data source* late Jurassic granites in the Jiaodong Peninsula (166–146 Ma; Zhang et al. 2010; Jiang et al. 2013; Ma et al. 2013; Zhao et al. 2016, 2018; Li et al. 2019), early Cretaceous granodiorites in the Jiaodong Peninsula (136–123 Ma; Zhang et al. 2010; Yang et al. 2012; Wang et al. 2014; Zhao et al. 2018; Li et al. 2019), and later early Cretaceous granites in the Jiaodong Peninsula (123–108 Ma; Huang et al. 2006; Goss et al. 2010; Zhang et al. 2010; Tang et al. 2014; Li et al. 2012, 2014; Yan and Shi 2016; Zhao et al. 2018; Gao et al. 2019; Li et al. 2019; Yan et al. 2019)

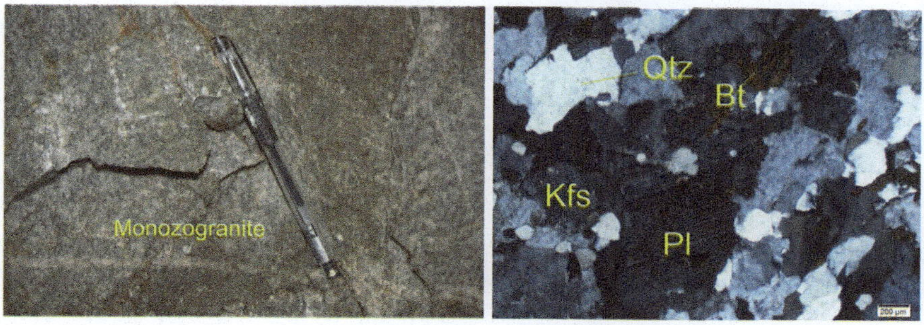

Fig. 3.35 Field photographs and microphotograph showing monzogranite (after Liang et al. 2020b)

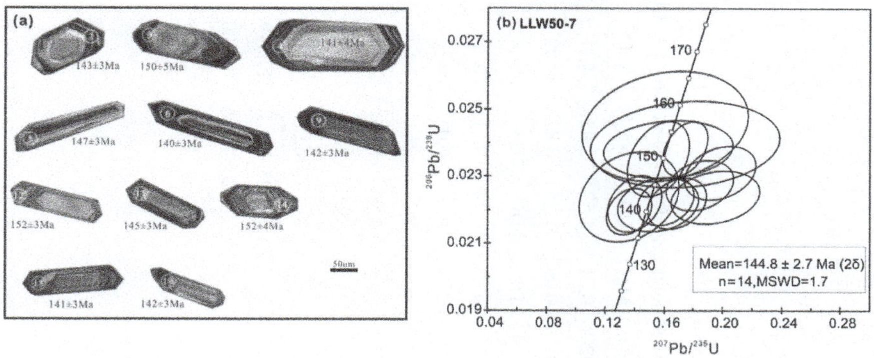

Fig. 3.36 **a** Cathodoluminescence (CL) images of zircons and **b** zircon U–Pb (**a**–**e**) concordia diagrams of the monzogranite (Data are from Liang et al. 2020a)

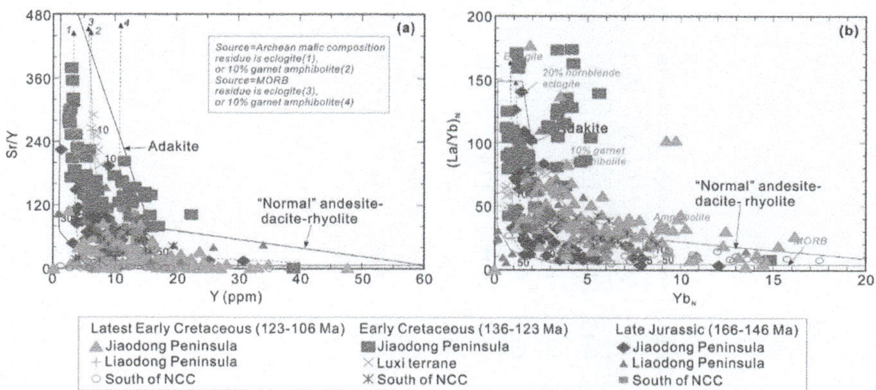

Fig. 3.37 Chondrite-normalized REE patterns (**a**, **c**, **e**) and primitive mantle-normalized immobile trace element patterns (**b**, **d**, **f**) for Late Mesozoic felsic rocks on the Jiaodong Peninsula. *Data source* Chondrite and primitive mantle values (Sun and McDonough 1989); The data and source of Late Mesozoic felsic rocks are listed in the Table 1

Table 3.12 Whole-rock geochemistry compositions of monzogranite from Linglong gold deposit (Data are from Liang et al. 2020b)

Sample	SiO_2	TiO_2	Al_2O_3	$^TFe_2O_3$	FeO	Fe_2O_3	MnO	MgO	CaO	Na_2O	K_2O	P_2O_5	LOI	Total	$Mg^{\#}$	
LLW50-7	74.09	0.15	14.32	1.26	0.8	0.37	0.02	0.21	1.52	3.92	4.11	0.03	0.42	99.97	25	
LLW50-8	75.31	0.07	13.73	0.68	0.4	0.24	0.01	0.09	1.36	3.87	4.09	0.01	0.6	99.78	21	
Sample	La	Ce	Pr	Nd	Sm	Eu	Gd	Tb	Dy	Ho	Er	Tm	Yb	Lu	Rb	Co
LLW50-7	23.6	46.6	5.18	17.4	2.42	0.61	1.69	0.19	0.74	0.11	0.29	0.03	0.18	0.03	95.6	0.84
LLW50-8	8.4	14.9	1.8	6.2	1	0.44	0.75	0.08	0.35	0.05	0.13	0.02	0.12	0.02	95.9	1.15
Sample	Ba	Th	U	Nb	Ta	K	Pb	Sr	P	Zr	Hf	Ti	Y	V	Cr	Ni
LLW50-7	2445	5.91	1.05	4.03	0.14	34118	37.5	528	128	87.6	44.61	928	2.8	14	7	0.81
LLW50-8	1470	3.02	0.84	1.96	0.11	33952	48.5	395	58	50.5	22.81	438	1.3	4	3	1.12

Sample	$^{87}Sr/^{86}Sr$	2δ	$(^{87}Sr/^{86}Sr)_i$	$^{143}Nd/^{144}Nd$	2δ	$\varepsilon_{Nd}(t)$	T (Ma)	$^{206}Pb/^{204}Pb$	$^{207}Pb/^{204}Pb$	$^{208}Pb/^{204}Pb$
LLW50-7	0.712055	6	0.710978	0.511504	7	−20.0	2563	17.119	15.503	37.987
LLW50-8	0.712486	5	0.711043	0.511438	9	−21.6	2684	17.100	15.502	37.907

Table 3.13 Whole-rock geochemical compositions of granodiorites from Linglong gold deposit (Data are from Liang et al. 2020b)

Sample	SiO_2	TiO_2	Al_2O_3	$^{T}Fe_2O_3$	FeO	Fe_2O_3	MnO	MgO	CaO	Na_2O	K_2O	P_2O_5	LOI	Total	$Mg^{\#}$	
LLW50-14	71.81	0.25	14.44	1.69	1.12	0.45	0.02	0.51	1.7	2.34	4.69	0.07	1.66	99.06	37	
LLW50-15	71.32	0.25	14.3	1.69	1.1	0.46	0.02	0.46	1.7	3.25	4.17	0.07	1.08	98.18	35	
LLW50-16	71.02	0.25	14.23	1.57	0.96	0.5	0.02	0.46	1.71	3.53	4.18	0.07	1.09	98.03	37	
LLW50-18	70.29	0.22	14.3	1.7	1.08	0.5	0.05	0.62	2.29	3.63	3.41	0.07	1.29	97.75	42	
Sample	La	Ce	Pr	Nd	Sm	Eu	Gd	Tb	Dy	Ho	Er	Tm	Yb	Lu	Rb	Ba
LLW50-14	50	86.2	8.89	28.5	3.91	0.77	2.73	0.31	1.29	0.22	0.56	0.08	0.51	0.08	132	1420
LLW50-15	50.9	89.2	9.14	29.2	3.9	0.76	2.78	0.31	1.31	0.21	0.57	0.09	0.5	0.08	111.4	1328
LLW50-16	49.9	85.7	8.82	27.9	3.73	0.75	2.69	0.3	1.24	0.21	0.56	0.08	0.46	0.08	112.4	1349
LLW50-18	27.8	49.3	5.23	17.3	2.5	0.67	1.83	0.22	0.96	0.15	0.4	0.06	0.38	0.06	91.3	1238
Sample	Th	U	Nb	Ta	K	Pb	Sr	P	Zr	Hf	Ti	Y	V	Cr	Co	Ni
LLW50-14	17.83	3.26	8.99	0.82	38933	26.2	196	296	128.1	6.14	1505	5.9	22	8	2.24	1.62
LLW50-15	18.59	2.94	7.99	0.81	34616	27.4	203	291	134.3	6.61	1489	6	24	8	2.28	1.61
LLW50-16	16.8	2.59	7.86	0.8	34699	23.6	186	290	113	5.69	1516	5.7	18	4	2.09	1.37
LLW50-18	7.55	1.87	6.57	0.78	28307	51.7	372	293	88.8	5.88	1348	4.2	25	13	2.24	4.67
Sample	$^{87}Sr/^{86}Sr$	2δ	$(^{87}Sr/^{86}Sr)_i$	$^{143}Nd/^{144}Nd$	2δ	$\varepsilon_{Nd}(t)$	T (Ma)	$^{206}Pb/^{204}Pb$	$^{207}Pb/^{204}Pb$	$^{208}Pb/^{204}Pb$						
LLW50-14	0.713100	3	0.710063	0.511454	9	−21.5	2653	17.039	15.455	38.056						
LLW50-15	0.713034	10	0.710560	0.511454	8	−21.5	2651	17.051	15.453	38.116						
LLW50-16	0.712993	6	0.710264	0.511436	8	−21.8	2680	17.032	15.449	38.090						
LLW50-18	0.710392	5	0.709283	0.511497	7	−20.7	2590	17.130	15.471	37.852						

The Guojialing granodiorites have high SiO_2 contents (62.4–77.7 wt%), and show LILE and LREE enrichments, HFSE depletion, negative $\varepsilon_{Nd}(t)$ (−22.3 to −11.5) and $\varepsilon_{Hf}(t)$ (−25.1 to −11.6) values, as well as high $\delta^{18}O$ values (7.46–8.67‰, Li et al. 2019), suggesting a crustal source. The $\varepsilon_{Hf}(t)$ values, which lie between the 1.9 and 2.5 Ga crustal evolution lines, imply a Palaeoproterozoic crustal source (Fig. 3.38). The T_{DM2} ages of $\varepsilon_{Nd}(t)$ and $\varepsilon_{Hf}(t)$ are 2931–2615 Ma and 3221–1917 Ma, respectively (Fig. 3.34). The Neoarchean and Palaeoproterozoic inherited zircon ages (2771–1617 Ma) from the Guojialing pluton were also consistent with ancient basement rocks in the NCC. Thus, the Guojialing pluton was likely sourced from partial melting of the Archean-Palaeoproterozoic continental crust in the NCC. Guojialing granodiorites also show adakitic characteristics, high Sr/Y and La/Yb ratios (Fig. 3.37), that similar to Late Jurassic granites, also suggesting a thickened lower crust source that are a plagioclase-free amphibolite–eclogite or a garnet–amphibolite crustal source.

In contrast, the Guojialing granitoids have higher $\varepsilon_{Nd}(t)$, $\varepsilon_{Hf}(t)$, and $\delta^{18}O$ values than those of the late Jurassic granites (Fig. 3.38), suggesting a change in the source. Yang et al. (2014) showed that the Guojialing granodiorite and dioritic enclaves occurred contemporaneously (zircon U-Pb age of 127 ~123 Ma), indicating a mixed magma source. Compared to the late Jurassic granites, the Guojialing granodiorites have a higher and wider range of $Mg^{\#}$ (23–71) values, approaching the values of the early Cretaceous lithospheric mantle-derived mafic dykes at Jiaodong ($Mg^{\#}$ = 55–78; Deng et al. 2017). The Guojialing granodiorites also show approximately similar values of $\varepsilon_{Nd}(t)$ and low $(^{87}Sr/^{86}Sr)_i$ values as the Jiaodong mafic dykes. These geochemical characteristics reveal a crustal source with added mantle-derived components. Wang et al. (2018) showed that the early Cretaceous mantle-derived mafic rocks (~129 Ma) in the Luxi terrane, also in the eastern NCC, have high zircon $\delta^{18}O$ values (7.7–9.1‰). In the Jiaodong Peninsula, the Guojialing granodiorites also have high $\delta^{18}O$ values (7.46–8.67‰, Li et al. 2019), which further supports the addition of mantle-derived material to the crustal source.

(3) **Latest Early Cretaceous granites (123–108 Ma)**

The latest Early Cretaceous intrusive rocks are composed of the Aishan complex (e.g. granodiorites, syenogranites, quartz syenogranites, alkali granites, and monzodiorites; 123–108 Ma), the Gushan granite (~120 Ma), and the Yashan coarse-grained monzogranite (116–113 Ma) in the Jiaobei terrane (Goss et al. 2010; Li et al. 2012; Zhao et al. 2018; Li et al. 2019); the Weideshan granite (~114 Ma), the Sanfoshan complex (e.g. monzonites, quartz monzonites, granodiorites, and granites; 118–114 Ma), the Haiyang K-feldspar granite (117–107 Ma), the Laoshan granitoids and syenogranite (123–116 Ma), and the Wulian-Dazhuangshan granite (122–109 Ma) in the Sulu Orogenic Belt (Huang et al. 2006; Goss et al. 2010; Zhang et al. 2010; Li et al. 2014; Tang et al. 2014; Yan and Shi 2016; Zhao et al. 2018; Gao et al. 2019;Yan et al. 2019). Four granodiorites (LLW50-14, −15, −16, and −18) collected samples from the −50-m drift of the Linglong gold deposit on the north-western Jiaodong Peninsula and displayed porphyritic textures (Fig. 3.39). Zircon from granodiorites

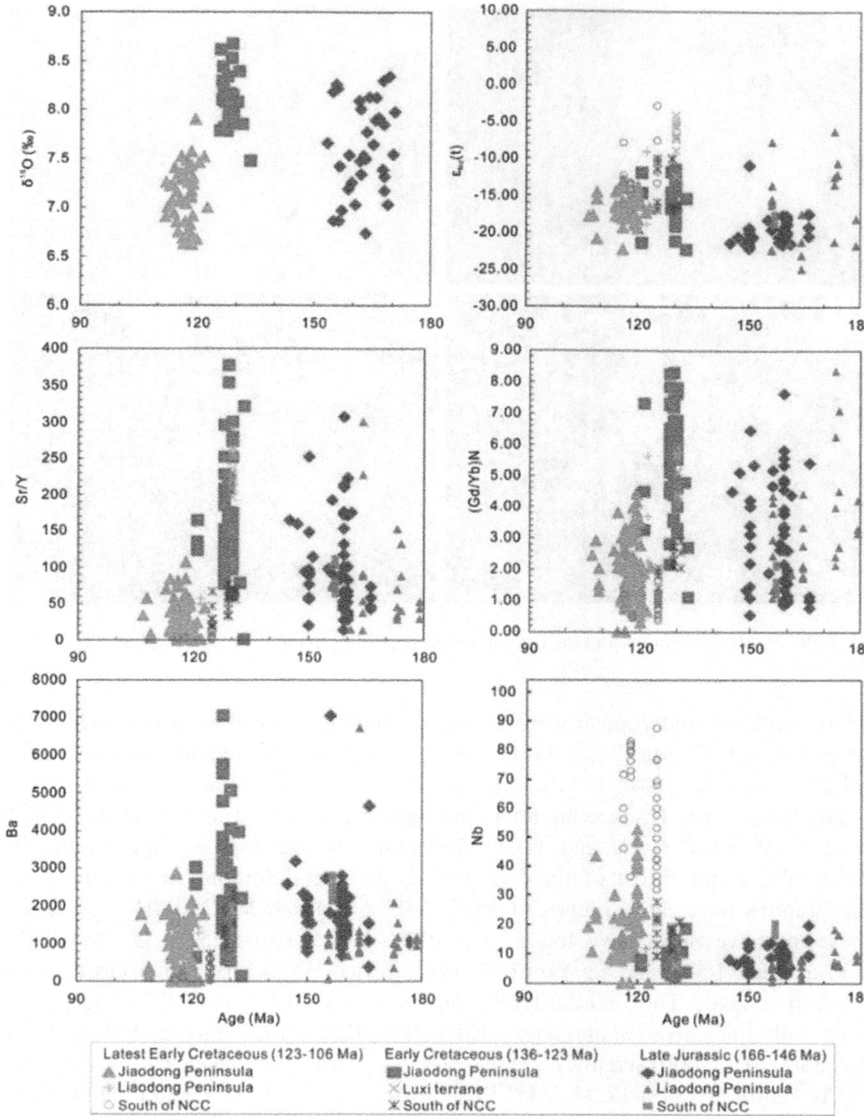

Fig. 3.38 Plots of variations in **a** $\delta^{18}O$, **b** $\varepsilon_{Nd}(t)$, **c** Sr/Y, **d** $(Gd/Yb)_N$, **e** Ba, and **f** Nb with time for Late Mesozoic felsic rocks on the Jiaodong Peninsula. Data source are listed Table 3.1

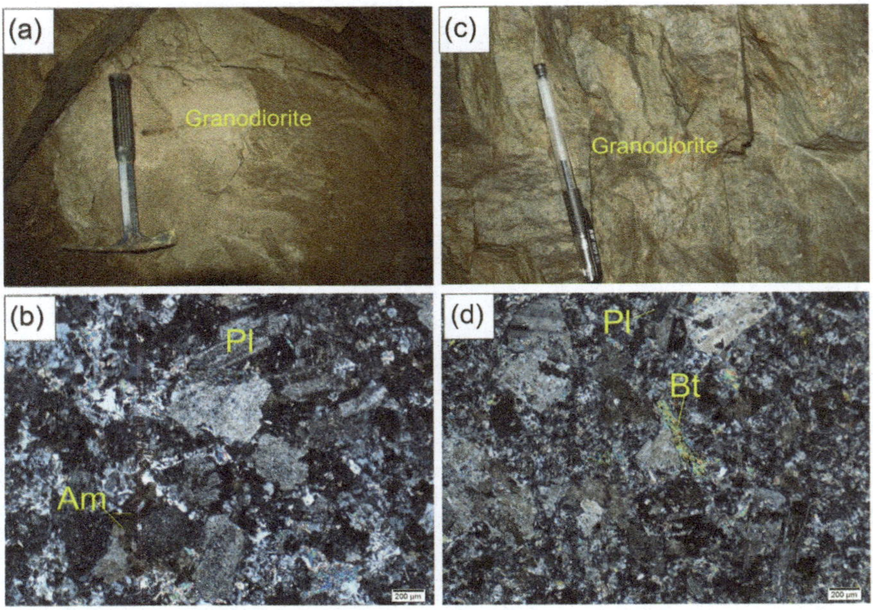

Fig. 3.39 Field photographs and microphotograph showing granodiorite (after Liang et al. 2020b)

in the Linglong gold deposit were chosen for dating and yielded a weighted mean age of 106.9 ± 2.3 Ma (2σ, MSWD = 1.4), including inherited ages of 1778 2031, and 2255 Ma (Fig. 3.40).

The latest Early Cretaceous felsic intrusive rocks (123–108 Ma) all have high contents of SiO_2, K_2O, and Na_2O, belonging to the high-K calc-alkaline to shoshonitic series. Most of these granitoids are metaluminous to peraluminous. The granites have large ranges of $Mg^{\#}$ (5.05–64.6) and MgO (0.04–3.87 wt%). These intrusive rocks have lower Sr/Y (0.27–108, average 25.3), Dy/Yb (0.39–2.49, average 1.54), and La/Yb (2.59–176, average 38.7) ratios than those of the two earlier stages. They exhibit HFSE depletion and LILE and LREE enrichments, along with a negative Eu anomaly (δEu = 0.02–0.78). They have high $(^{87}Sr/^{86}Sr)_i$ (0.705400–0.724600) and low $\varepsilon_{Nd}(t)$ (−22.3 to −3) values, and have narrow ranges of $(^{206}Pb/^{204}Pb)_i$ (16.119–17.342), $(^{207}Pb/^{204}Pb)_i$ (15.386–15.505), and $(^{208}Pb/^{204}Pb)_i$ (36.194–37.930) isotopic ratios. Zircons exhibit $\varepsilon_{Hf}(t)$ values of −25.5 to −5.4, with T_{DM2} ages of 3900–1515 Ma (Figs. 3.29, 3.30, 3.31, 3.32, 3.33 and 3.34).

All the latest early Cretaceous granites likely originated from a continental crustal source, based on their characteristic of low $\varepsilon_{Nd}(t)$ (−22.3 to −3) and zircon $\varepsilon_{Hf}(t)$ (−25.5 to −5.4) values, high $\delta^{18}O$ values (6.6 to 7.9‰), enrichment in LREEs and LILEs, and depletion in HFSEs. Most of the T_{DM2} ages derived from Hf isotopes varied from 2500–1515 Ma, with a few >3000 Ma (Fig. 3.34) ages, indicating that the latest early Cretaceous granites originated from the partial melting of the Palaeoproterozoic crustal source and minor Neoarchean components, which is consistent with the composition of ancient basement of the Jiaodong Peninsula (Jahn et al.

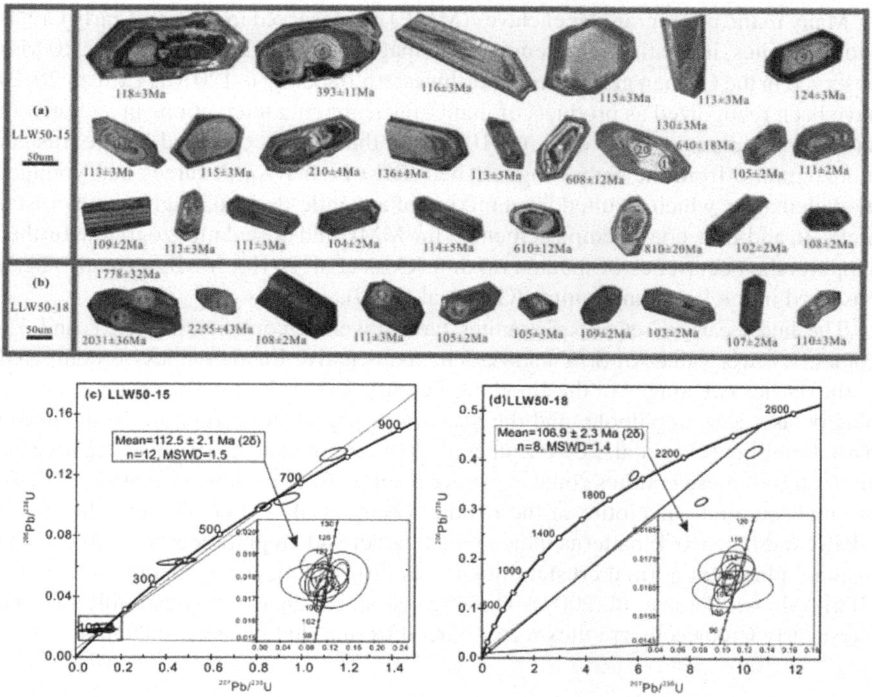

Fig. 3.40 a–b Cathodoluminescence (CL) images of zircons and **c–d** zircon U–Pbconcordia diagrams of the granodiorites (Data are from Liang et al. 2020b)

2008). However, the NCC's ancient lower crust is characterized by low Rb content, enrichment in Hf, and depletion in Zr, Th, and U (Gao et al. 1998), which is the opposite of the latest early Cretaceous granites. This shows that these components are unlikely to have been directly or solely sourced from partial melting of the ancient NCC lower crust. Isotopically, these granites have similar $(^{87}Sr/^{86}Sr)_i$ ratios as the lower NCC crust, but their $\varepsilon_{Nd}(t)$ values plot between the ranges of the lower crust and lithospheric mantle-derived mafic dykes at Jiaodong. The zircon $\varepsilon_{Hf}(t)$ values of these granites vary from −24.6 to −13.5, implying the presence of enriched sources such as sediments or an enriched mantle with/without depleted mantle components (Woodhead et al. 2001). The latest early Cretaceous granites show similar $\delta^{18}O$ values (6.6 to 7.9‰ from the Aishan felsic complex (Li et al. 2019); 5.65–7.78‰ and 4.68–7.08‰ from the Haiyang felsic complex, Li et al. 2014) to contemporary mafic rocks (~115 Ma) from the Jiaodong Peninsula (zircon $\delta^{18}O$ values of 5.4–6.7‰; Guo et al. 2014). Moreover, these granites have far lower O isotope components than sediments, and thus have additional mantle-derived materials in the source with minor or no sediment input. Liu et al. (2019) also assumed that felsic-intermediate rocks (<200 Ma) in the NCC were the result of crustal re-working without a significant addition of depleted mantle melt.

Many mafic microgranular enclaves (MME) are observed in the latest early Cretaceous granites, indicating the occurrence of magma mixing. MME (118.5 ± 1.0 Ma) observed in the Gushan granite from northwestern Jiaodong (~120 Ma, Li et al. 2012) have been recognized as products of mafic–felsic magma interactions in a relatively late magmatic stage. Goss et al. (2010) found that MME, captured by the Yashan monzogranite from the Sulu orogenic belt, have higher $\varepsilon_{Hf}(t)$ values and younger modelled ages, which resulted from mixing of a mantle-derived magma with crustal melts. In addition, coeval emplacement of the MME and hosted monzogranite further supports the occurrence of magma mixing (Goss et al. 2010). MME have also been observed in the Laoshan granites (Gao et al. 2019).

The latest early Cretaceous granites have lower Sr contents, higher Y and Yb contents, lower values of Sr/Y and La/Yb, and negative Eu anomalies, as compared to the earlier two stages in the Jiaodong Peninsula (Fig. 3.37). This indicates more plagioclase, less amphibole, and the absence of garnet in residual minerals during partial melting (Gao et al. 2014; Zou et al. 2019; Zeng et al. 2019). The negative Ba anomalies of these granites could also have resulted from the presence of plagioclase or small amounts of biotite in the residue (Zeng et al. 2019). The enrichment of HREE and flat HREE patterns indicate that garnet and amphibole were stable as the residual phases at normal crustal thicknesses (20–40 km, corresponding to 0.8–1.2 GPa) (Alonso-Perez et al. 2009). These geochemical characteristics imply that the latest early Cretaceous granites were sourced from a shallower crustal depth.

3.7.1.2 The Variation in the Thickness of Lower Crust of the Jiaodong Peninsula

Late Jurassic to early Cretaceous intrusive rocks in the Jiaodong Peninsula present a geologic record of thickening and thinning of the lower crust and the destruction of the eastern NCC. As mentioned previously, there are large variations in the major element, trace element, and radiogenic isotope compositions of these granites from 166–108 Ma (Fig. 3.38). Firstly, a series of dramatic changes offer a window into the temporal changes in the late Mesozoic lower crust beneath the Jiaodong Peninsula, because three stages of granite emplacement originating from partial melting of the LCC. Late Jurassic granites originated from partial melting of the eastern NCC Archean lower crust. The early Cretaceous granodiorites were derived from partial melting of the eastern NCC Archean-Palaeoproterozoic lower crust with the addition of mantle-derived materials. However, the latest early Cretaceous granites were derived from mixing of lower crustal-origin felsic magma and mantle-derived mafic melts. Secondly, whole rock Sr/Y and La/Yb ratios of rocks in subduction-related arcs have been proved to be correlate well with crustal thickness at global and regional scales (Chapman et al. 2015; Profeta et al. 2015) and the changes in crustal thickness during collisional process (Hu et al. 2017). But, Jiaodong granites display extremely high Sr/Y ratios (>100) that are greater than differentiation within the thickest crust based on Chapman et al. (2015). Zhu et al. (2017b) has applied the whole rock La/Yb ratios correlation from Profeta et al. (2015) to track changes of crustal thickness in

Tibet as a consequence of rollback and then breakoff of the subducting Neo-Tethyan slab, in which some rocks show very high Sr/Y ratios (>100). In this study, we also used La/Yb ratios that based on Profeta et al. (2015) to track the crustal thickness of the Jiaodong Peninsula, eastern NCC, which has undergone the subduction-related metasomatism during the Late Mesozoic. The variations in La/Yb (or Sm/Yb and Gd/Yb) ratios from the late Jurassic to the latest early Cretaceous granites (Fig. 3.41) indicate a thickening of the LCC during the late Jurassic, continued thickening of the LCC in the early Cretaceous, and thinning of the LCC in the latest early Cretaceous. Thirdly, these changes in the magma source also reflect regional changes in tectonic stress. The early stage (166–144 Ma) granites are mainly typical I-type granites, but the latter two stages (136–123 Ma and 123–108 Ma) shift to A-type and/or highly fractionated I-type granites. This change suggests that the regional tectonic stress that affects the eastern NCC has altered from compression to an intraplate extension after ~144 Ma.

(1) **Thickened lower continental crust caused by the subduction of Palaeo–Pacific Oceanic lithosphere**

Late Jurassic granites (166–146 Ma) in the Jiaodong Peninsula show high Sr/Y and La/Yb ratios, indicating a thickened lower crustal source (>40 km) (Fig. 3.37). At the southern margin of the eastern NCC, the Wuzhangshan felsic pluton (157–156 Ma) also has high SiO_2, Al_2O_3, and Sr contents, low Y and Yb contents, with high Sr/Y (16–44.5) and La/Yb (16.1–31.8) ratios (Zou et al. 2019). In the Liaodong Peninsula in the northeastern NCC, late Jurassic granites (179–156 Ma) also have similar adakitic characteristics (e.g. high Sr/Y = 33–112 and La/Yb = 10.3–110.3) (Wu et al. 2005). Moreover, the Late Triassic Longtou–Chaxinzi–Xiaoweishahe and Nankouqian–Xidadingzi granitoids in the northern Liaodong Peninsula, which have

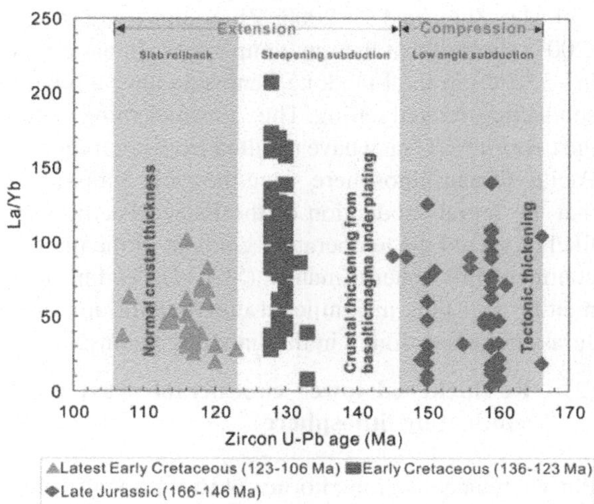

Fig. 3.41 Plot of changes in La/Yb ratios in felsic intrusive rocks over time for the Jiaodong Peninsula, eastern NCC. *Data source* are listed Table 3.1

been recognized to be originated from partial melting of ancient lower crustal materials with involvement of mantle components (Yang et al. 2012), characterized by strong negative and variable whole rock $\varepsilon_{Nd}(t)$ and $\varepsilon_{Hf}(t)$ and zircon $\varepsilon_{Hf}(t)$ values, however, lack of adakitic features. This implied that the Late Jurassic lower crust of the eastern NCC has been changed and thickened. It has been shown that late Jurassic magmatism in the eastern NCC likely occurred in (1) a post-collisional extensional setting following the collision of the NCC and the Yangtze craton (YC) (Zhang et al. 2010; Jiang et al. 2012) or (2) a compressional setting resulting from the subduction of the Palaeo–Pacific Ocean lithosphere (Wu et al. 2005; Ma et al. 2013). We can exclude the former mechanism due to the following reasons: (1) the subduction and collision of the NCC and the YC (240–200 Ma) took place prior to late Jurassic magmatism in Jiaodong for approximately 100–80 Ma (Ye et al. 2000; Zheng et al. 2003); (2) the large difference between Sr–Nd isotopic compositions of late Triassic crust-derived syenites (~215 Ma) in the Sulu orogenic belt and late Jurassic granites at Jiaodong, indicates that the source changed (Fig. 3.8); and (3) late Jurassic Linglong granites in the Jiaodong Peninsula are predominantly typical I-type granites, indicating a compressional setting rather than a post-collisional extensional setting. Moreover, both Jurassic granitoids from the Liaodong Peninsula and from the southern margin of the NCC, are also typical I-type granites (Wu et al. 2005; Zou et al. 2019). Therefore, late Jurassic thickened lower crust (>40 km) and associated magmatism in the eastern NCC did not have a direct genetic relationship with the Triassic collision of the NCC and YC. Although the Neoproterozoic inherited zircons (600–800 Ma) appeared in the low-Mg adakitic felsic intrusions and likely originated from the YC because this craton inherited much of the Neoproterozoic magmatism (750–600 Ma) (Chen and Jahn 1998; Zheng et al. 2006), crustal modification derived from YC subduction during 240–200 Ma (Liang et al. 2020a).

Previous studies have shown that the Palaeo–Pacific Oceanic lithosphere subducted westward under the Asian continent since the early Jurassic (ca. 200–190 Ma; Zhou and Li 2000; Zhou and Long 2017; Tang et al. 2018). Wu et al. (2007) showed that the low-temperature Jurassic granitoids (formation temperature of ~750°C) in the Liaodong Peninsula have a relationship with fluids, indicating a subduction-related setting. Thus, the thickening of middle-late Jurassic lower crust in the eastern NCC may have resulted from compression and subduction of the Palaeo-Pacific Ocean lithosphere since the early Jurassic. Liu et al. (2019) has suggested that the initial subduction of the Palaeo-Pacific Ocean lithosphere at a low angle likely lowered the temperature gradient of the mantle wedge and the overlying sub-continent lithospheric mantle (SCLM), leading to limited asthenospheric mantle melting and thus insignificant crust-mantle mixing, which is consistent with late Jurassic granites found in the Jiaodong Peninsula.

(2) **Re-thickened lower continental crust resulted from steepening of the subducting lithosphere**

Early Cretaceous granodiorites (136–123 Ma) from the Jiaodong Peninsula are A-type and/or highly fractionated I-type granites, suggesting an extensional tectonic environment. The contemporary metamorphic core complexes (MCCs), pull-apart basins, A-type granites, and dyke swarms occurred in the eastern NCC, showing a regional extensional tectonic environment. The early Cretaceous A-type alkaline granites (131–118 Ma) have also been observed in the Liaodong Peninsula and on the southern margin of the NCC (132–125 Ma) (Wu et al. 2005; Yang et al. 2007; Zou et al. 2019). Early Cretaceous pull-apart basins (140–120 Ma), including the Songliao Basin, Jiaolai Basin, and Hefei Basin (Wei et al. 2010; Charles et al. 2013), have also occurred. Early Cretaceous MCCs (135–115 Ma) developed in the eastern NCC, including the Liaonan, Yunmengshan, and Hohhot MCCs on the northern margin, and the Xiaoqinling MCC on the southern margin (Zhu et al. 2015). Therefore, the eastern NCC became a regional extensional setting ~136 Ma ago.

Moreover, early Cretaceous granodiorites show higher Sr and lower Y contents with higher Sr/Y ratios and La/Yb ratios than the late Jurassic granites (Figs. 3.37 and 3.41), implying thicker crust. Because La/Yb (or Sm/Yb and Gd/Yb) ratios show a positive relationship with crustal thickness (Profeta et al. 2015; Farner and Lee 2017; Zhu et al. 2017; Zeng et al. 2019), the higher La/Yb values from the early Cretaceous granodiorites, compared to those from the late Jurassic granites, further indicate crustal thickening after 146 Ma. This has also been observed on the southern margin of the eastern NCC, such as in the Huashani complex (ca. 132–125 Ma) with higher Sr/Y ratios (17.6–76.8) and La/Yb ratios (18.5–43.2) than those of the Wuzhangshan pluton (ca. 157–156 Ma) (Zou et al. 2019). Numerical modeling and geophysical studies reveal that slab steepening (a precursor of rollback) and rollback are important in producing topographic uplift (e.g., Göğüş 2015; Smith et al. 2017; Zeng et al. 2019). In particular, if the temperature distribution in the lithospheric mantle is heterogeneous during rollback, then crustal uplift can occur in the inboard region due to the removal of lithospheric mantle (or arc root), whereas crustal subsidence will develop in the trenchward region (Göğüş 2015; Cassel et al. 2018; Zeng et al. 2019). The crustal thickening of the arc likely resulted from basaltic magma underplating caused by slab rollback and slab breakoff, such as in the Gangdese arc at 70–45 Ma (Zhu et al. 2017b). The Guojialing granodiorites (136–123 Ma) in the Jiaodong Peninsula have also shown interaction between lithospheric mantle-derived melts and the source. Wu et al. (2007) has shown that the early Cretaceous (131–118 Ma) high-temperature granitoids in the Liaodong Peninsula are likely related to mantle plume turbulence in the mantle and decompression partial melting of the upwelling asthenosphere.

Niu et al. (2003) has proposed that the oldest age of the Zhangjiakou Formation (135 Ma) in the eastern NCC represents the time when tectonic stress altered from compression to extension. Liu et al. (2017) assumed that trench retreat of the western Pacific occurred at 137–130 Ma. Dai et al. (2016) studied Mesozoic–Cenozoic mafic igneous rocks and assumed that the subduction angle of the Palaeo-Pacific plate changed from flat to steep at ~135 Ma. Zheng et al. (2018) thought that the Palaeo-Pacific Ocean lithosphere initially subducted westward at a low angle (flat slab subduction), and then began to sink or rollback to allow asthenospheric upwelling at ~144 Ma. A higher angle of subduction would lead to decoupling between the subducting Palaeo-Pacific slab and the mantle wedge, which would result in not only rollback of the downgoing slab, but also upwelling of asthenospheric mantle, generating mantle-derived melts (Dai et al. 2016). The occurrence of early Cretaceous lithospheric mantle-derived mafic dykes (144–143 Ma) in the Zichuan area of the eastern NCC suggests the presence of mantle thermal anomalies (Liu et al. 2008).

Thus, we assume that a steepening of the subducting Palaeo-Pacific Oceanic lithosphere took place since ~144 Ma, and that this was the first-step in the process of slab rollback, which then triggered basaltic magma underplating in the region and the subsequent crustal thickening of the Jiaodong Peninsula.

(3) **Thinning of lower continental crust induced by slab rollback**

The late stage granites (123–108 Ma) in the Jiaodong Peninsula, characterized by low Sr/Y and La/Yb ratios (Fig. 3.37), were derived from thinner crustal depths (20–40 km) than those of the earlier two stages of granites. The thinning of the LCC source for late granites (126–110 Ma) also took place at the southern margin of the eastern NCC. Zou et al. (2019) showed that the latest early Cretaceous granitoids (ca. 125–110 Ma) in the southern margin of the NCC have low Sr contents and high Y contents, with low Sr/Y (0.25–13.2) and La/Yb (5.48–57.8) ratios, suggesting they have a thinned crustal source (<33 km). Gao and Zhao (2017) concluded that the granites at 126–112 Ma in the southern margin of the NCC also belong to A-type and/or highly fractionated I-type granites with low contents of Sr, δEu values, and Sr/Y ratios, thus forming in an intracontinental extensional environment.

In fact, lithospheric extension should occur after rollback of the oceanic lithosphere (Göğüş 2015 and references therein), and if this had occurred, the driving force for thrusting in the NCC lithosphere should have been removed. In this study, we have found that the eastern NCC was a dominantly regional intracontinental extensional environment since the early Cretaceous, which is likely relevant to the rollback of the Palaeo-Pacific plate since ~144 Ma. Dai et al. (2016) showed that higher angle subduction of the western Pacific plate would induce not only rollback of the downgoing slab, but also asthenospheric mantle convection beneath the retreating slab. Intense asthenospheric mantle convection likely eroded the weakened overlying lithospheric mantle, generating a larger scale of lithospheric mantle-derived melts, and resulting in thinning of the lithosphere. Meanwhile, decompression and melting of upwelling asthenospheric mantle also occurred, presenting as the asthenospheric mantle-derived mafic dykes observed in the southwestern Jiaodong Peninsula (122–121 Ma; Ma et al. 2014). Previous studies have shown that the large-scale of lithospheric mantle-derived mafic dykes formed at 130–110 Ma with a peak at ~123 Ma over the entire Jiaodong Peninsula (Deng et al. 2017). Afterward, upwelling and underplating of the mantle-derived magma led to partial melting of the lower crust, producing massive volumes of felsic intrusive rocks at 123–108 Ma. The stress regime during this period was dominated by extension, generating metamorphic core complexes, pull-apart basins, A-type granites, and dyke swarms. Therefore, the rollback of the subducting Paleo-Pacific Ocean lithosphere took place since ~123 Ma.

3.7.2 *The Thinning of Lithospheric Mantle Beneath Jiaodong Peninsula*

Geochronology studies have shown that from 130 Ma to 110 Ma, large-scale emplacement events of mafic dykes (lamprophyre, diabase, gabbro, etc.) occurred in the Jiaodong Peninsula. Two types of mafic dyke rocks have been identified: those derived from the lithospheric mantle (rc-like mafic dykes enriched in LILEs and LREEs, but depleted in HFSEs; 110–130 Ma) and those derived from the asthenospheric mantle (OIB-like mafic dykes, no HFSEs depletion and LREEs enrichment; ~122 Ma) (Guo et al. 2004; Yang et al. 2004; Tan et al. 2007; Liu et al. 2009; Cai et al. 2013, 2015; Ma et al. 2014, 2016). In addition, because the regular only partially melts when the thickness of the lithosphere is less than ~80 km (McKenzie and Bickle 1988). The development of OIB-like mafic dykes derived from the asthenospheric mantle in the Jiaodong area shows that the thickness of the lithosphere is less than 80 km (Ma et al. 2014). In addition, the arc-like mafic dyke rocks are widely distributed in the Jiaodong Peninsula, whereas the determined OIB-like mafic dykes are only distributed in the Jiaojia gold deposit and surrounding areas in the north-western part of the Jiaodong Peninsula (Ma et al. 2014). Therefore, the widely developed arc-like mafic dykes and the appearance of local OIB-like mafic dykes indicate that the Mesozoic lithospheric thinning event of the North China Craton is also an important geological event in the Jiaodong Peninsula.

Regarding the thinning mechanism of the lithosphere of the North China Craton, existing interpretations include thermo-mechanical erosion (Xu 2001, 2014), lithospheric detachment (Wu et al. 2005), and lower crustal delamination (Gao et al. 2004, 2004; Liu et al. 2008a, 2008b, 2009). The thermo-mechanical erosion model shows that the lower part of the lithosphere is gradually eroded by asthenospheric materials, in a slow process that lasts for at least 100 million years (Meng et al. 2007). At the beginning of the lithospheric thinning, influence of magmatism on the thinning of the lithosphere was characterised as the dominant factor in thermodynamic erosion. When the lithosphere becomes thinner than ~80 km, the asthenosphere melts partially, generating magma derived from the asthenosphere mantle (Xu et al. 2004; Xu 2014). In the model of lithospheric detachment, it is speculated that the thick bottom layer of the lithospheric mantle suddenly detached to the asthenosphere, marking a short event lasting ~10 million years (Menzies et al. 2007). The rapid and short-term detachment sets of magmatism which causes generation of mantle-derived magma in the lithosphere and asthenosphere, as well as of mixed felsic magma derived from the crust or mantle (Deng et al. 2007; Wu et al. 2008). The lower crustal delamination model implies that the lower crustal delamination led to the expansion of the lithosphere and upwelling of the asthenosphere. This event started at about 159 Ma and continued into the Early Cretaceous (Gao et al. 2004; Liu et al. 2009). Subsequently, the melting of the delaminated eclogite produced a violent reaction between the silicon-rich melt and the mantle peridotite melt. The main source of magma is the lithospheric mantle hybridised by reduced pressure melting (Gao et al. 2008; Liu et al. 2009) (Fig. 3.42).

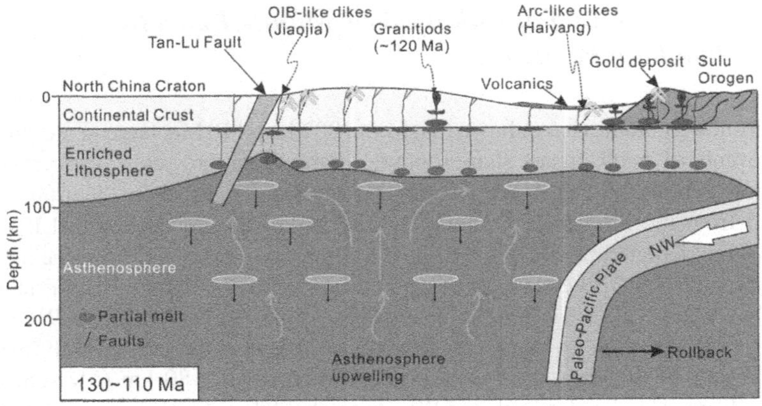

Fig. 3.42 A possible model showing the generation of mafic dykes in Jiaodong Peninsula (after Liang et al. 2017)

In the early Cretaceous period (140–120 Ma), the formation of the Jiaolai Basin was caused by the right strike-slip extension of the Tanlu Fault Zone (Zhang et al. 2003). Early Cretaceous magmatic rocks in the Jiaodong Peninsula have the characteristics of ancient crustal sources, and due to the influence of the mantle, plutonic intrusions such as Taidingding granite and granodiorite of Guojialing, Gushan, and Shijiu (Yang et al. 2005; Li et al. 2012), indicating that the lower crust has undergone mantle-derived magma underplating (Tang et al. 2014). In the early Cretaceous, arc-like mafic dykes intruded extensively the Jiaodong Peninsula for 110 to 130 Ma, indicating that the thinning of the lithosphere in the eastern North China Craton was a long-term process. The magmatism that occurred in the Jiaodong Peninsula in the Early Cretaceous may be related to the upwelling of the asthenosphere and the expansion of the lithosphere caused by the thinning of the lithosphere (Yang et al. 2005). The thinning may be related to the Paleo-Pacific Ocean Plate subduction (Wu et al. 2005; Zhu et al. 2011) or/and involving a plate retraction (Yang et al. 2007; Jiang et al. 2010). The Jiaodong mafic dyke originate from an enriched lithospheric mantle that has undergone metasomatism from the subduction of the Paleo-Pacific Plate. In this chapter, we have provided evidence that the lithospheric mantle of Jiaodong experienced metasomatism of the subduction-related fluids of the Paleo-Pacific Plate (including oceanic sediments) during the Mesozoic. Starting from about 160 Ma, the retreat of the Paleo-Pacific subducted slab and related trench movement led to the extension of the overlying lithosphere, an asthenospheric, rise, which together caused continental rifting and magmatic activity (Wu et al. 2007; Jiang et al. 2010; Mao et al. 2011; Ma et al. 2014). The slab retreat led to extensional deformation and thinning of the lithosphere. As the bottom of the lithospheric mantle was continuously thermally eroded by the asthenosphere, partial melting occurred, forming the Huaziyu lamprophyre dyke in the north-eastern part of the North China Craton (~155 Ma), and Mengyin mafic dyke rocks (~144 Ma) and Zichuan mafic dyke rocks (~143 Ma) in western Shandong Province. At the same time, the mantle-derived magma undercut and partially melted the crust to form the Jurassic low-Mg adakitic monzogranite (~145 Ma). After 135 Ma, the Izanaki Plate changed its subduction direction in the NE direction. Its converging direction was approximately parallel to the continental edge of eastern Asia. This subduction caused a strong tension, disconnection, and left-strike-slip deformation in the eastern part of the North China Craton (Mao et al. 2011). The continuous slab retreat and left-lateral strike-slip structure of the Izanaki oceanic lithosphere led to the formation of continental rifts in the eastern part of the North China Craton (Sun et al. 2007), which further led to the delamination of the lithospheric mantle basement in the eastern part of the North China Craton. Lithospheric delamination caused upwelling of the asthenosphere, causing partial melting of the lithospheric mantle to form the parent magma of the Jiaodong mafic dykes (high-Mg clinopyroxene, namely IA and IB clinopyroxene rims and II group clinopyroxene). Subsequently, the mantle-derived magma underplated to form the new lower crust (IA clinopyroxene core), and at the same time caused partial melting of the ancient lower crust to form an andesitic–dacite melt (IB clinopyroxene core).

References

Adam, J., and T.H. Green. 1994. The effects of pressure and temperature on the partitioning of Ti, Sr and REE between amphibole, clinopyroxene and basanitic melts. *Chemical Geology* 117: 219–233.

Alonso-Perez, R., O. Müntener, and P. Ulmer. 2009. Magmatic garnet and amphibole fractionation in the roots of island arcs: Experimental constraints on andesitic liquids. *Contributions to Mineralogy and Petrology* 157: 541–558.

Asimow, P.D., J.E. Dixon, and C.H. Langmuir. 2004. A hydrous melting and fractionation model for mid-ocean ridge basalts: application to the Mid-Atlantic Ridge near the Azores. *Geochemistry, Geophysics, Geosystems* 5: 1–24.

Batki, A., P. Elemér, D. Gábor, and S. Alasdair. 2014. Petrogenetic significance of ocellar camptonite dykes in the Ditrău Alkaline Massif, Romania: Lithos, 200–201, 181–196.

Bell, D.R., P.D. Ihinger, and G.R. Rossman. 1995. Quantitative analysis of trace OH in garnet and pyroxenes. *American Mineralogist* 80: 465–474.

Ben, O.D., W.M. White, and J. Patchett. 1989. The geochemistry of marine sediments, island arc magma genesis, and crust-mantle recycling: Earth Planet. *Science Letters* 94: 1–21.

Blundy, J.D., T.J. Falloon, B.J. Wood, and J.A. Dalton. 1995. Sodium partitioning between clinopyroxene and silicate melts. *Journal of Geophysical Researches* 100: 15501–15515.

Cai, Y.C., H.R. Fan, M. Santosh, F.F. Hu, K.F. Yang, and Z. Hu. 2015. Subduction-related metasomatism of the lithospheric mantle beneath the southeastern north china craton: Evidence from mafic to intermediate dykes in the northern sulu orogen. *Tectonophysics* 659: 137–151.

Cai, Y.C., H.R. Fan, M. Santosh, X. Liu, F.F. Hu, K.F. Yang, T.G. Lan, Y.H. Yang, and Y.S. Liu. 2013. Evolution of the lithospheric mantle beneath the southeastern North China Craton: Constraints from mafic dykes in the Jiaobei terrain. *Gondwana Research* 24: 601–621.

Callegaro, S., Andreamarzoli, H. Bertrand, C. Massimo, L. Reisberg, Christine., B. Giuliano, E.W. Robert, and R. Merle. 2013. Upper and lower crust recycling in the source of CAMP basaltic dykes from southeastern North America: Earth Planet. *Science Letters* 376: 186–199.

Castillo, P.R. 2012. Adakite petrogenesis. *Lithos* 134 (3): 304–316.

Castillo, P.R., P.E. Janney, and R.U. Olidum. 1999. Petrology and geochemistry of Camiguin Island, southern Philippines: Insights to the source of adakites and other lavas in a complex arc setting. *Contributions to Mineralogy and Petrology* 134(1): 33–51.

Chapman, J.B., M.N. Ducea, L. Profeta, and P.G. DeCelles. 2015. Tracking changes in crustal thickness during orogenic evolution with Sr/Y; an example from the Western U.S. Cordillera. *Geology* 43: 919–923.
Charles, N., R. Augier, C. Gumiaux, P. Monié, Y. Chen, M. Faure, and R. Zhu. 2013. Timing, duration and role of magmatism in wide rift systems: insights from the Jiaodong Peninsula (China, East Asia). *Gondwana Research* 24: 412–428.
Chen, S., S.Y. O'Reilly, X. Zhou, W.L. Griffin, G. Zhang, M. Sun, and M. Zhang. 2001. Thermal and petrological structure of the lithosphere beneath Hannuoba, Sino-Korean Craton, China: Evidence from xenoliths. *Lithos* 56: 267–301.
Chiaradia, M., D. Merino, and R. Spikings. 2009. Rapid transition to long-lived deep crustal magmatic maturation and the formation of giant porphyry-related mineralization (Yanacocha, Peru). *Earth and Planetary Science Letters* 288: 505–515.
Chung, S.L., D. Liu, J. Ji, M.F. Chu, H.Y. Lee, D.J. Wen, C.-H. Lo, T.Y. Lee, Q. Qian, and Q. Zhang. 2003. Adakites from continental collision zones: melting of thickened lower crust beneath southern Tibet. *Geology* 31 (11): 1021–1024.
Dai, L.Q., Z.F. Zhao, Y.F. Zheng, Y.J. An, and F. Zheng. 2017. Geochemical Distinction between Carbonate and Silicate Metasomatism in Generating the Mantle Sources of Alkali Basalts. *Journal of Petrology* 58: 863–884.
Dai, L.Q., Y.F. Zheng, and Z.F. Zhao. 2016. Termination time of peak decratonization in North China: Geochemical evidence from mafic igneous rocks. *Lithos* 240: 327–336.
Defant, M.J., and M.S. Drummond. 1990. Derivation of some modern arc magmas by melting of young subducted lithosphere. *Nature* 347: 662–665.
Deng, J., X.F. Liu, Q.F. Wang, and R.G. Pan. 2015. Origin of the Jiaodong-type Xinli gold deposit, Jiaodong Peninsula, China: Constraints from fluid inclusion and C-D–O–S–Sr isotope compositions: Ore Geol. *Rev.* 65: 674–686.
Deng, J., X.F. Liu, Q.F. Wang, Y. Dilek, and Y.Y. Liang. 2017. Isotopic characterization and petrogenetic modeling of the early cretaceous mafic Diking-Lithospheric extension in the north china craton, Eastern Asia. *Geological Society of America Bulletin* 129: 1379–1407.
Deng, J., Q.F. Wang, G.J. Li, and M. Santosh. 2014. Cenozoic tectono-magmatic and metallogenic processes in the Sanjiang region, southwestern China. *Earth-Science Reviews*, 268–299.
Deng, J.F., S.G. Su, Y.L. Niu, C. Liu, G.C. Zhao, X.G. Zhao, S. Zhou, and Z.X. Wu. 2007. A possible model for the lithospheric thinning of North China Craton: evidence from the Yanshanian (Jura-Cretaceous) magmatism and tectonism: Lithos 96, 22–35.
Doucet, L.S., A.H. Peslier, D.A. Ionov, A.D. Brandon, A.V. Golovin, A.G. Goncharov, and I.V. Ashchepkov. 2014. High water contents in the Siberian cratonic mantle linked to metasomatism: An FTIR study of Udachnaya peridotite xenoliths. *Geochimica et Cosmochimica Acta* 137: 159–187.
Duggen, S., K. Hoernle, P. Bogaard, and D. Garbe-Schonberg. 2005. Post-collisional transition from subduction-to intraplatetype magmatism in the westernmost Mediterranean: Evidence for continental-edge delamination of subcontinental lithosphere. *Journal of Petrology* 46: 1155–1201.
Elliott, T., T. Plank, A. Zindler, W. White, and B. Bourdon. 1997. Element transport from slab to volcanic front at the Mariana arc. *Journal Geophysical Research* 102: 14991–15019.
Ernst, R.E., J.W. Head, E. Parfitt, E. Grosfils, and L. Wilson. 1995. Giant radiating dyke swarms on Earth and Venus. *Earth-Science Reviews* 39: 1–58.
Farner, M.J., and C.T.A. Lee. 2017. Effects of crustal thickness on magmatic differentiation in subduction zone volcanism: A global study. *Earth and Planetary Science Letters* 470: 96–107.
Fischer, T.P. 2008. Fluxes of volatiles (H_2O, CO_2, N_2, Cl, F) from arc volcanoes. *Geochemical Journal* 42: 21–38.
Furman, T., and D. Graham. 1999. Erosion of lithospheric mantle beneath the East African Rift system: Geochemical evidence from the Kivu volcanic province. *Lithos* 48: 237–262.
Gao, S., R.L. Rudnick, W. Xu, H. Yuan, Y. Liu, R.J. Walker, I.S. Puchtel, X. Liu, H. Huang, X. Wang, and J. Yang. 2008. Recycling deep cratonic lithosphere and generation of intraplate magmatism in the North China Craton. *Earth and Planetary Science Letters* 270: 41–53.

Gao, X.Y., T.P. Zhao, Z.W. Bao, and A.Y. Yang. 2014. Petrogenesis of the Early Cretaceous intermediate and felsic intrusions at the southern margin of the North China Craton: Implications for crust-mantle interaction. *Lithos* 206: 65–78.

Gao, Y., Y. Niu, M. Duan, Q. Xue, P. Sun, S. Chen, and Y. Chen. 2019. The petrogenesis and tectonic significance of the Early Cretaceous intraplate granites in eastern China: The Laoshan granite as an example. *Lithos* 328: 200–211.

Göğüş, O.H. 2015. Rifting and subsidence following lithospheric removal in continental back arcs. *Geology* 43: 3–6.

Goss, S.C., S.A. Wilde, F.Y. Wu, and J.H. Yang. 2010. The age, isotopic signature and significance of the youngest Mesozoic granitoids in the Jiaodong Terrane, Shandong Province, North China Craton. *Lithos* 120: 309–326.

Griffin, W.L., N.J. Pearson, E. Belousova, S.E. Jackson, E.V. Achterbergh, S.Y. O'Reilly, and S.R. Shee. 2000. The Hf isotope composition of cratonic mantle: LAM-MC-ICPMS analysis of zircon megacrysts in kimberlites: Geochim. *Cosmochim. Acta* 64: 133–147.

Guo, F., W.M. Fan, Y.J. Wang, M. Zhang. 2004. Origin of early Cretaceous calc-alkaline lamprophyres from the Sulu orogen in eastern China: implications for enrichment processes beneath continental collisional belt. *Lithos* 78: 291–305.

Guo, F., W.M. Fan, C.W. Li, C.Y. Wang, H.X. Li, L. Zhao, and J.Y. Li. 2014. Hf-Nd-O isotopic evidence for melting of recycled sediments beneath the Sulu Orogen, North China. *Chemical Geology* 381: 243–258.

Guo, F., E. Nakamuru, W.M. Fan, K. Kobayoshi, and C.W. Li. 2007. Generation of Palaeocene adakitic andesites by magma mixing, Yanji Area, NE China. *Journal of Petrology* 80: 1–32.

Halls, H.C. 1982. The importance and potential of mafic dyke swarms in studies of geodynamic processes. *Geoscience Canada* 9: 145–154.

He, L. 2014. Numerical modeling of convective erosion and peridotite–melt interaction in big mantle wedge: Implications for the destruction of the North China Craton. *Journal of Geophysical Research: Solid Earth* 119: 2013J–10657J.

Hermann, J., and D. Rubatto. 2009. Accessory phase control on the trace element signature of sediment melts in subduction zones. *Chemical Geology* 265: 512–526.

Hofmann, A.W. 2003. Sampling mantle heterogeneity through oceanic basalts: isotopes and trace elements. In *The Mantle and Core. Treatise on Geochemistry* ed by Carlson R.W.,. ElservierPergamon, Oxford, pp. 61–101.

Hou, G.T., M. Santosh, X.L. Qian, G.S. Lister, and J.H. Li. 2008. Configuration of the Late Paleoproterozoic supercontinent Columbia: Insights from radiating mafic dyke swarms. *Gondwana Research* 14: 395–409.

Hu, F., M.N. Ducea, S. Liu, and J.B. Chapman. 2017. Quantifying crustal thickness in continental collisional belts: global perspective and a geologic application. *Scientific Reports* 7 (1): 1–10.

Huang, X.L., J.W. Zhong, and Y.G. Xu. 2012. Two tales of the continental lithospheric mantle prior to the destruction of the North China Craton: Insights from Early Cretaceous mafic intrusions in western Shandong, east China: Geochim. *Cosmochim. Acta* 96: 193–214.

Huang, X.L., Y.G. Xu, and D.Y. Liu. 2004. Geochronology, petrology and geochemistry of the granulite xenoliths from Nushan, east China: Implication for a heterogeneous lower crust beneath the Sino-Korean Craton. *Geochimica et Cosmochimica Acta* 68: 127–149.

Ionov, D.A., W.L. Griffin, and S.Y. O'Reilly. 1997. Volatile-bearing minerals and lithophile trace elements in the upper mantle: Chem. *Geology* 141: 153–184.

Jahn, B.M., D. Liu, Y. Wan, B. Song, and J. Wu. 2008. Archean crustal evolution of the Jiaodong Peninsula, China, as revealed by zircon SHRIMP geochronology, elemental and Nd-isotope geochemistry. *American Journal of Science* 308 (3): 232–269.

Jiang, N., J. Chen, J. Guo, and G. Chang. 2012. In situ zircon U-Pb, oxygen and hafnium isotopic compositions of jurassic granites from the North China Craton: Evidence for Triassic subduction of continental crust and subsequent metamorphism-related ^{18}O depletion. *Lithos* 142–143: 84–94.

Jiang, P., K.F. Yang, H.R. Fan, X. Liu, Y.C. Cai, and Y.H. Yang. 2016. Titanite-scale insights into multi-stage magma mixing in Early Cretaceous of NW Jiaodong terrane, North China Craton. *Lithos* 258: 197–214.

Jiang, Y.H., S.Y. Jiang, H.F. Ling, and P. Ni. 2010. Petrogenesis and tectonic implications of Late Jurassic shoshonitic lamprophyre dikes from the Liaodong Peninsula, NE China: Miner. *Petrol* 100: 127–151.

Kawamoto, T., M. Yoshikawa, Y. Kumagai, Ma. H.T. Mirabueno, M. Okuno, and T. Kobayashi. 2013. Mantle wedge infiltrated with saline fluids from dehydration and decarbonation of subducting slab. *Proceedings of the National Academy of Sciences* 110: 9663–9668.

Kelemen, P.B., J.K. Hanghø, and A.R. Greene. 2003. One View of the Geochemistry of Subduction-related Magmatic Arcs, with an Emphasis on Primitive Andesite and Lower Crust. *Treatise on Geochemistry* 3: 593–659.

Keller, C.B., B. Schoene, M. Barboni, K.M. Samperton, and J.M. Husson. 2015. Volcanic-plutonic parity and the differentiation of the continental crust. *Nature* 523: 301–307.

Khan, P.K., S. Shamim, M. Mohanty, P. Kumar, and J. Banerjee. 2017. Myanmar-Andaman-Sumatra subduction margin revisited: Insights of arc-specific deformations. *Journal of Earth Science* 28: 683–694.

Kogiso, T., and M.M. Hirschmann. 2001. Experimental study of clinopyroxenite partial melting and the origin of ultra-calcic melt inclusions. *Contributions to Mineralogy and Petrology* 142: 347–360.

Kogiso, T., Y. Tatsumi, and S. Nakano. 1997. Trace element transport during dehydration processes in the subducted oceanic crust: 1. Experiments and implications for the origin of ocean island basalts. *Earth and Planetary Science Letters* 148: 193–205.

Kovács, I., J. Hermann, H.S.C. O'Neill, J.F. Gerald, M. Sambridge, and G. Horvath. 2008. Quantitative absorbance spectroscopy with unpolarized light: Part II. Experimental evaluation and development of a protocol for quantitative analysis of mineral IR spectra. *American Mineralogist* 93: 765–778.

Krüger, J.C., L.H. Romerr, and H. Kämpf. 2013. Late Cretaceous ultramafic lamprophyres and carbonatites from the Delitzsch Complex, Germany: Chem. *Geology* 353: 140–150.

Li, H., M.X. Ling, X. Ding, H. Zhang, C.Y. Li, D.Y. Liu, and W.D. Sun. 2014. The geochemical characteristics of Haiyang A-type granite complex in Shandong, eastern China. *Lithos* 200–201: 142–156.

Li, X.H., Z.X. Li, W.X. Li, X.C. Wang, and Y. Gao. 2013. Revisiting the "C-type adakites" of the Lower Yangtze River Belt, central eastern China: in-situ zircon Hf–O isotope and geochemical constraints. *Chemical Geology* 345: 1–15.

Li, X.C., H.R. Fan, M. Santosh, F.F. Hu, K.F. Yang, T.G. Lan, Y.S. Liu, and Y.H. Yang. 2012. An evolving magma chamber within extending lithosphere: An integrated geochemical, isotopic and zircon U-Pb geochronological study of the Gushan granite, eastern North China Craton. *Journal of Asian Earth Sciences* 50: 27–43.

Li, X.H., H.R. Fan, F.F. Hu, P. Hollings, K.F. Yang, and X. Liu. 2019. Linking lithospheric thinning and magmatic evolution of Late Jurassic to Early Cretaceous granitoids in the Jiaobei Terrane, southeastern North China Craton. *Lithos* 324: 280–296.

Li, Z.X., and X.H. Li. 2007. Formation of the 1300-km-wide intracontinental orogen and postorogenic magmatic province in Mesozoic South China: A flat-slab subduction model. *Geology* 35: 179–182.

Liang, Y.Y., J. Deng, X.F. Liu, Q.F. Wang, Y. Ma, T.X. Gao, and L.H. Zhao. 2019. Water contents of early cretaceous mafic dikes in the Jiaodong Peninsula, Eastern North China craton: Insights into an enriched lithospheric mantle source metasomatized by Paleo–Pacific plate subduction–related fluids. *Journal of Geology* 127: 262–343.

Liang, Y.Y., J. Deng, X.F. Liu, Q.F. Wang, C. Qin, Y. Li, Y. Yang, and M. Zhou. 2018. Major and trace element, and Sr isotope compositions of clinopyroxene phenocrysts in mafic dykes on Jiaodong Peninsula, southeastern North China Craton: Insights into magma mixing and source metasomatism. *Lithos* 302–303: 480–495.

Liang, Y.Y., W.H. Guo, Y. Ma, and E.Q. Zhao. 2020a. Geochemical evidence for thickening and thinning of lower crust beneath the jiaodong peninsula: Implications for late mesozoic destruction of the Eastern North China Craton. *Lithosphere*. https://doi.org/10.2113/2020/8831664.

Liang, Y.Y., X.F. Liu, C. Qin, Y. Li, J. Chen, J.Y. Jiang. 2017. Petrogenesis of early cretaceous mafic dikes in southeastern Jiaolai basin, Jiaodong Peninsula China. *International Geology Review* 59: 131–150.

Liang, Y.Y., X.F. Liu, Q.F. Wang, R. Zhao, and Y. Ma. 2020b. Late Mesozoic magmatism in the Jiaodong Peninsula, East China: Implications for crust–mantle interactions and lithospheric thinning of the eastern North China Craton. *Geoscience Frontiers* 11: 895–914.

Liang, Y.Y., J. Deng, X.F. Liu, Q.F. Wang, Y. Ma, T.X. Gao, E.Q. Zhao, and Z.H. Zhou. 2020c. Geochronology and geochemistry of the Palaeoproterozoic mafic dikes in the Jiaobei terrane: implications for tectonic evolution of the Jiao-Liao-Ji Belt, eastern North China Craton. *International Geology Review*. https://doi.org/10.1080/00206814.2020.1746932.

Liu, X.F., J. Deng, Y.Y. Liang, Q.F. Wang, G.J. Li, Y. Ma, L.J. Xu, Y.H. Lu. 2020. Geochemical, mineralogical and chronological studies of mafic intermediate dykes in the Jiaodong Peninsula: implications for Late Mesozoic mantle source metasomatism and lithospheric thinning of the eastern North China Craton. *International Geology Review* 62: 2239–2260.

Liu, D., Z. Zhidan, Z. Di-Cheng, N. Yaoling, D.J. Depaolo, T., M.H., M. Xuanxue, D. Guochen, S. Zhou, G.S. Chen, C.Z. Zhao, and J.L. Liu. 2014. Postcollisional potassic and ultrapotassic rocks in southern Tibet: Mantle and crustal origins in response to India–Asia collision and convergence: Geochim. *Cosmochim. Acta* 143: 207–231.

Liu, J., Q.K. Xia, E. Deloule, J. Ingrin, H. Chen, and M. Feng. 2015. Water content and oxygen isotopic composition of alkali basalts from the taihang mountains, china: Recycled oceanic components in the mantle source. *Journal of Petrology* 56: 681–702.

Liu, P.H., F.L. Liu, C.H. Liu, F. Wang, J.H. Liu, H. Yang, J. Cai, and J.R. Shi. 2013. Petrogenesis, P-T-t path, and tectonic significance of high-pressure mafic granulites from the Jiaobei terrane, North China Craton. *Precambrian Research* 233: 237–258.

Liu, S., R. Hu, S. Gao, C. Feng, B. Yu, G. Feng, Y. Qi, T. Wang, and I.M. Coulson. 2009. Petrogenesis of Late Mesozoic mafic dykes in the Jiaodong Peninsula, eastern North China Craton and implications for the foundering of lower crust: Lithos 113, 621–639.

Liu, S., R.Z. Hu, S. Gao, C.X. Feng, L. Qi, H. Zhong, T. Xiao, Y.Q. Qi, T. Wang, and I.M. Coulson. 2008. Zircon U-Pb geochronology and major, trace elemental and Sr–Nd–Pb isotopic geochemistry of mafic dykes in western Shandong Province, east China: constrains on their petrogenesis and geodynamic significance. *Chemical Geology* 38: 329–345.

Liu, S.C., Q.K. Xia, S.H. Choi, E. Deloule, P. Li, and J. Liu. 2016. Continuous supply of recycled Pacific oceanic materials in the source of Cenozoic basalts in SE China: the Zhejiang case. *Contributions to Mineralogy and Petrology* 171: 100.

Liu, X.F., Q.F. Wang, Q.Z. Zhang, S.J. Yang, Y.U. Liang, Y. Zhang, Y. Li, and T. Guan. 2017. Genesis of the Permian karstic Pingguo bauxite deposit, western Guangxi China. *Mineralium Deposita* 7: 1–18.

Liu, Y.S., S. Gao, Z.C. Hu, C.G. Gao, K.Q. Zong, and D.B. Wang. 2010. Continental and oceanic crust recycling-induced meltperidotite interactions in the Trans-North China orogen: U-Pb dating, hf isotopes and trace elements in zircons from mantle xenoliths. *Journal of Petrology* 51: 537–571.

Liu, Y.S., S. Gao, S.Y. Jin, S.H. Hu, M. Sun, and Z.B. Zhao. 2001. Geochemistry of lower crustal xenoliths from Neogene Hannuoba basalt, North China Craton: Implications for petrogenesis and lower crustal composition. *Geochimica et Cosmochimica Acta* 65: 2589–2604.

Liu, Y.S., S. Gao, C.A. Lee, S.H. Hu, X.M. Liu, and H.L. Yuan. 2005. Melt-peridotite interactions: links between garnet pyroxenite and high-$Mg^{\#}$ signature of continental crust. *Earth and Planetary Science Letters* 234: 39–57.

Lloyd, F.E. 1981. Upper-mantle metasomatism beneath a continental rift: clinopyroxenes in alkaline mafic lavas and nodules from South-West Uganda. *Mineralogical Magazine* 44: 315–323.

Long, X., S.A. Wilde, Q. Wang, C. Yuan, X.C. Wang, J. Li, Z.Q. Jiang, and W. Dan. 2015. Partial melting of thickened continental crust in central Tibet: Evidence from geochemistry

and geochronology of Eocene adakitic rhyolites in the northern Qiangtang Terrane. *Earth and Planetary Science Letters* 414: 30–44.
Ma, L., S.Y. Jiang, and A. Hofman. 2014. Lithospheric and asthenospheric sources of lamprophyres in the Jiaodong Peninsula: A consequence of rapid lithospheric thinning beneath the North China Craton? *Geochimica et Cosmochimica Acta* 124: 250–271.
Ma, L., S.Y. Jiang, A.W. Hofman, Y.G. Xu, B.Z. Dai, and M.L. Hou. 2016a. Rapid lithospheric thinning of the North China Craton: New evidence from cretaceous mafic dykes in the Jiaodong Peninsula. *Chemical Geology* 432: 1–15.
Ma, L., S.Y. Jiang, B.Z. Dai, Y.H. Jiang, M.L. Hou, W. Pu, and B. Xu. 2013. Multiple sources for the origin of Late Jurassic Linglong adakitic granite in the Shandong Peninsula, eastern China: zircon U-Pb geochronological, geochemical and Sr–Nd–Hf isotopic evidence. *Lithos* 162: 251–263.
Ma, Q., Y.G. Xu, J.P. Zheng, W.L. Griffin, L.B. Hong, and L. Ma. 2016b. Coexisting early Cretaceous high-Mg andesites and adakitic rocks in the North China Craton: The role of water in intraplate magmatism and cratonic destruction. *Journal of Petrology* 57: 1279–1308.
Mao, J.W., F. Pirajno, and N. Cook. 2011. Mesozoic metallogeny in East China and corresponding geodynamic settings—An introduction to the special issue: Ore Geology Reviews 43, 1–7.
Martin, H., R.H. Smithies, R. Rapp, J.F. Moyen, and D. Champion. 2005. An overview of adakite, tonalite-trondhjemite-granodiorite (TTG), and sanukitoid: relationships and some implications for crustal evolution. *Lithos* 79: 1–24.
Mckenzie, D.P., and M.J. Bickle. 1988. The volume and composition of melt generated by extension of the lithosphere. *Journal of Petrology* 29: 625–679.
Menzies, M.A., Y.G. Xu, H.F. Zhang, and W.M. Fan. 2007. Integration of geology, geophysics and geochemistry: A key to understanding the North China Craton: Lithos 96, 1–21.
Monecke, T., P. Dulski, and U. Kempe. 2007. Origin of convex tetrads in rare earth element patterns of hydrothermally altered siliceous igneous rocks from the Zinnwald Sn–W deposit Germany. *Geochimica et Cosmochimica Acta* 71 (2): 335–353.
Morimoto, N., J. Fabries, A.K. Ferguson, I.V. Ginzburg, M. Ross, F.A. Seifert, J. Zussman, K. Aoki, and G. Gottardi. 1988. Nomenclature of pyroxenes. *American Mineralogist* 73: 1123–1133.
Müntener, O., P.B. Keleman, and T.L. Grove. 2001. The role of H_2O during crystallization of primitive arc magmas under the uppermost mantle conditions and genesis of igneous pyroxenites: An experimental study. *Contributions to Mineralogy and Petrology* 141: 643–658.
Nakamura, E., and I. Kushiro. 1998. Trace element diffusion in jadeite and diopside melts at high pressures and its geochemical implication. *Geochimica et Cosmochimica Acta* 62: 3151–3160.
Pang, C.J., X.C. Wang, Y.G. Xu, S.N. Wen, Y.S. Kuang, and L.B. Hong. 2015. Pyroxenite-derived Early Cretaceous lavas in the Liaodong Peninsula: Implication for metasomatism and thinning of the lithospheric mantle beneath North China Craton. *Lithos* 227: 77–93.
Perugini, D., C.P. De Campos, D.B. Dingwell, M. Petrelli, and G. Poli. 2008. Trace element mobility during magma mixing: preliminary experimental results. *Chemical Geology* 256: 146–157.
Pilet, S., J. Hernandez, and B. Villemant. 2002. Evidence for high silicic melt circulation and metasomatic events in the mantle beneath alkaline provinces: the Na-Fe-augitic green-core pyroxenes in the tertiary alkali basalts of the Cantal massif (French Massif Central). *Mineralogy and Petrology* 76: 39–62.
Plank, T., K.A. Kelley, M.M. Zimmer, E.H. Hauri, and P.J. Wallace. 2013. Why do mafic arc magmas contain 4 wt% water on average? Earth Planet. *Science Letters* 364: 168–179.
Plank, T., and C.H. Langmuir. 1998. The chemical composition of subducting sediment and its consequences for the crust and mantle. *Chemical Geology* 145: 325–394.
Profeta, L., M.N. Ducea, J.B. Chapman, S.R. Paterson, S.M.H. Gonzales, M. Kirsch, L. Petrescu, and P.G. DeCelles. 2015. Quantifying crustal thickness over time in magmatic arcs. *Scientifc Reports* 5: 17786.
Putirka, K., H. Mikaelian, F. Ryerson, and H. Shaw. 2003. New clinopyroxene-liquid thermobarometers for mafic, evolved and volatile-bearing lava compositions, with applications to lavas from Tibet and the Snake River Plain. *American Mineralogist* 88: 1542–1554.

Rapp, R.P., N. Shimizu, and M.D. Norman. 2003. Growth of early continental crust by partial melting of eclogite. *Nature* 425: 605–609.

Richards, J.P., and R. Kerrich. 2007. Special paper: Adakite-like rocks: their diverse origins and questionable role in metallogenesis. *Economic Geology* 102: 537–576.

Rivalenti, G., R. Vannucci, E. Rampone, M. Mazzucchelli, G.B. Piccardo, E.M. Piccirillo, and L. Ottolini. 1996. Peridotite clinopyroxene chemistry reflects mantle processes rather than continental versus oceanic settings. *Earth and Planetary Science Letters* 139: 423–437.

Rock, N.M. 1987. The nature and origin of lamprophyres: an overview. *Geological Society, London, Special Publications* 30: 191–226.

Rudnick, R.L., and S. Gao. 2003. Composition of the continental crust. *Treatise on geochemistry* 3: 659.

Skjerlie, K.P., and A.E. Patino Douce. 2002. The fluid-absent partial melting of a zoisite-bearing quartz eclogite from 10 to 32 GPa; Implications for melting in thickened continental crust and for subduction-zone processes. *Journal of Petrology* 43: 291–314.

Smith, M.E., E.J. Cassel, B.R. Jicha, B.S. Singer, and A.S. Canada. 2017. Hinterland drainage closure and lake formation in response to middle Eocene Farallon slab removal Nevada, USA. *Earth and Planetary Science Letters* 479: 156–169.

Spandler, C., and C. Pirard. 2013. Element recycling from subducting slabs to arc crust: A review. *Lithos* 170–171: 208–223.

Streck, M.J., W.P. Leeman, and J. Chesley. 2007. High-magnesian andesite from Mount Shasta: A product of magma mixing and contamination, not a primitive mantle melt. *Geology* 35: 351–354.

Sun, W.D., X. Ding, Y.H. Hu, and X.H. Li. 2007. The golden transformation of the Cretaceous plate subduction in the west Pacific. *Earth and Planetary Science Letters* 262: 533–542.

Sun, J.G., S.X. Hu, H.F. Ling, and Y. Ye. 2000. Study on element geochemistry and diagenesis of high-potassium-potassium dyke rocks in two types of gold ore fields in northwestern Jiaozhou. *Geochimica* 29: 143–152.

Tahmasbi, Z., M. Kiani, and A.A. Khalaji. 2016. Petrology and geochemistry of diabasic dikes and andesitic-basaltic lavas in Noorabad-Harsin ophiolite, SE of Kermanshah Iran. *Journal of Earth Science* 27: 935–944.

Tan, J., J.H. Wei, Y.G. Li, W.J. Tan, G. Dazhao, and C. Yang. 2007. Geochemical Characteristics of Late Mesozoic Dikes, Jiaodong Peninsula, North China Craton: Petrogenesis and Geodynamic Setting: Int. *Geology Reviews* 49: 931–946.

Tan, J., J.H. Wei, Y.J. Li, L.B. Fu, H.M. Li, W.J. Shi, and N. Tian. 2015. Origin and geodynamic significance of fault-hosted massive sulfide gold deposits from the Guocheng-Liaoshang metallogenic belt, eastern Jiaodong Peninsula: Rb–Sr dating, and H-O–S–Pb isotopic constraints. *Ore Geology Reviews* 65: 687–700.

Tan, J., J.H. Wei, W.J. Shi, B. Feng, Y.J. Li, and L.B. Fu. 2013. Origin of dyke swarms by mixing of metasomatized subcontinental lithospheric mantle-derived and lower crustal magmas in the Guocheng fault belt, Jiaodong Peninsula, North China Craton. *Geological Journal* 48: 516–530.

Tang, H.Y., J.P. Zheng, and C.M. Yu. 2008. Age and composition of the Rushan intrusive complex in the northern Sulu orogen, eastern China: petrogenesis and lithospheric mantle evolution: Geol. *Mag.* 1: 1–17.

Tang, L., and M. Santosh. 2018. Neoarchean-Paleoproterozoic terrane assembly and Wilson cycle in the North China Craton: an overview from the central segment of the Trans-North China Orogen. *Earth-Science Reviews* 182: 1–27.

Tatsumi, Y. 2000. Continental crust formation by crustal delamination in subduction zones and complementary accumulation of the enriched mantle I component in the mantle. *Geochemistry, Geophysics, Geosystems* 1: 2000G–2094G.

Thirlwall, M.F., T.E. Smith, A.M. Graham, N. Theodorou, P. Hollings, J.P. Davidson, and R.J. Arculus. 1994. High field strength element anomalies in arc lavas: Source or process?. *Journal of Petrology* 35: 819–838.

Tsuchiya, N., S. Suzuki, J.I. Kimura, and H. Kagami. 2005. Evidence for slab melt/mantle reaction: petrogenesis of Early Cretaceous and Eocene high-Mg andesites from the Kitakami Mountains, Japan. *Lithos* 79: 179–206.

Turner, S., N. Arnaud, J. Liu, N. Rogers, C. Hawkesworth, N. Harris, S. Kelley, P. Van Calsteren, and W. Deng. 1996. Post-collision, shoshonitic volcanism on the Tibetan Plateau: Implications for convective thinning of the lithosphere and the source of ocean island basalts. *Journal of Petrology* 37: 45–71.

Wade, J.A., T. Plank, E.H. Hauri, K.A. Kelley, K. Roggensack, and M. Zimmer. 2008. Prediction of magmatic water contents via measurement of H_2O in clinopyroxene phenocrysts. *Geology* 36: 799–802.

Wallace, P.J. 2005. Volatiles in subduction zone magmas: concentrations and fluxes based on melt inclusion and volcanic gas data. *Journal of Volcanology and Geothermal Research* 140: 217–240.

Wang, C.M., Y. Lu, X. He, M. Jessell, Q. Wang, and J. Zhang. 2016a. The Paleoproterozoic diorite dykes in the southern margin of the North China Craton: Insight into rift-related magmatism. *Precambrian Research* 277: 26–46.

Wang, H., S. Wang, Z. Xu, B. Fu, Z. Zhao, Z. Li, and J. Li. 2018. Geochemical and Sr–Nd–Pb–Hf–O isotopic compositions of the Tiezhai complex: Implications for lithosphere destruction of the North China Craton. *Gondwana Research* 61: 203–221.

Wang, H., Z. Xu, X. Lu, B. Fu, J. Lu, X. Yang, and Z. Zhao. 2016b. Two-types of Early Cretaceous adakitic porphyries from the Luxi terrane, eastern North China Block: Melting of subducted Paleo-Pacificslab and delaminated newly underplated lower crust. *Lithos* 240: 140–154.

Wang, Q.F., J. Deng, X.F. Liu, R. Zhao, and S.H. Cai. 2016c. Provenance of late carboniferous bauxite deposits in the North China Craton: New constraints on marginal arc construction and accretion processes. *Gondwana Research* 38: 86–98.

Wang, Z.L., L.Q. Yang, J. Deng, M. Santosh, H.F. Zhang, Y. Liu, R.H. Li, T. Huang, X.L. Zheng, and H. Zhao. 2014. Gold-hosting high Ba-Sr granitoids in the Xincheng gold deposit, Jiaodong Peninsula, East China: Petrogenesis and tectonic setting. *Journal of Asian Earth Sciences* 95: 274–299.

Wei, H.H., J.L. Liu, and Q.R. Meng. 2010. Structural and sedimentary evolution of the southern Songliao basin, northeast China and implications for hydrocarbon prospectivity. *AAPG Bulletin* 94: 533–566.

Woodhead, J.D., J.M. Hergt, J.P. Davidson, and S.M. Eggins. 2001. Hafnium isotope evidence for 'conservative' elemen mobility during subduction zone processes: Earth Planet. *Science Letters* 192: 331–346.

Woods, S.C., S. Mackwell, and D. Dyar. 2000. Hydrogen in diopside: diffusion profles. *American Mineralogist* 85: 480–487.

Woolley, A.R., S.C. Bergman, A.D. Edgar, M.J. Le, Bas, R.H. Mitchell, N.M.S. Rock, and B.H. Scott Smith. 1996.Classification of lamprophyres, lamproites, kimberlites, and the kalsilitic, melilitic, and leucitic rocks: Can. *Mineral* 34: 175–186.

Workman, R.K., and S.R. Hart. 2005. Major and trace element composition of the depleted MORB mantle (DMM). *Earth and Planetary Science Letters* 231: 53–72.

Wu, F.Y., J.Q. Lin, S.A. Wilde, X.O. Zhang, and J.H. Yang. 2005. Nature and significance of the early cretaceous giant igneous event in Eastern China. *Earth and Planetary Science Letters* 233: 103–119.

Wu, F.Y., Y.G. Xu, S. Gao, and J.P. Zheng. 2008. Lithospheric thinning and destruction of North China Craton: Acta Petrol. *Sin.* 24: 1145–1174. (in Chinese with English abstract).

Wu, F.Y., Y.H. Yang, L.W. Xie, J.H. Yang, and P. Xu. 2006. Hf isotopic compositions of the standard zircons and baddeleyites used in U-Pb geochronology: Chem. *Geology* 234: 105–126.

Xia, Q.K., J. Liu, S.C. Liu, I. Kovács, M. Feng, and L. Dang. 2013. High water content in mesozoic primitive basalts of the North China Craton and implications on the destruction of cratonic mantle lithosphere. *Earth Planetary Science Letters* 361: 85–97.

Xia, Q.K., Y.M. Sheng, X.Z. Yang, and H.M. Yu. 2005. Heterogeneity of water in garnets from UHP eclogites, eastern Dabieshan, China. *Chemical Geology* 224: 237–246.

Xia, Q.K., X.Z. Yang, E. Deloule, Y.M. Sheng, and Y.T. Hao. 2006. Water in the lower crustal granulite xenoliths from Nushan, eastern China. *Journal of Geophysical Research: Solid Earth* 111: B11202.

Xie, S.W., Y.B. Wu, Z.M. Zhang, Y.C. Qin, X.C. Liu, H. Wang, Z.W. Qin, Q. Liu, and S.H. Yang. 2012. U-Pb ages and trace elements of detrital zircons from early cretaceous sedimentary rocks in the Jiaolai basin, north margin of the sulu UHP terrane: Provenances and tectonic implications. *Lithos* 154: 346–360.

Xu, H.J., C.Q. Ma, Y.R. Song, J.F. Zhang, and K. Ye. 2012a. Early Cretaceous intermediate-mafic dykes in the Dabie orogen, eastern China: Petrogenesis and implications for crust–mantle interaction. *Lithos* 154: 83–99.

Xu, J.F., R. Shinjo, M.J. Defant, Q. Wang, and R.P. Rapp. 2002. Origin of Mesozoic adakitic intrusive rocks in the Ningzhen area of east China: PARTIAL melting of delaminated lower continental crust? *Geology* 30: 1111–1114.

Xu, R., Y.S. Liu, X.R. Tong, Z.C. Hu, K.Q. Zong, and S. Gao. 2013a. In-situ trace elements and Li and Sr isotopes in peridotite xenoliths from Kuandian, North China Craton: Insights into Pacific slab subduction-related mantle modification. *Chemical Geology* 354: 107–123.

Xu, W.L., Q.J. Zhou, F.P. Pei, D.B. Yang, S. Gao, Q.L. Li, and Y.H. Yang. 2013b. Destruction of the North China Craton: Delamination or thermal/chemical erosion? Mineral chemistry and oxygen isotope insights from websterite xenoliths. *Gondwana Research* 23: 119–129.

Xu, Y.G. 2001. Thermo-tectonic destruction of the Archaean lithospheric keel beneath the Sino-Korean Craton in China: Evidence, timing and mechanism. *Physics and Chemistry of the Earth, Part A: Solid Earth and Geodesy* 26: 747–757.

Xu, Y.G. 2014. Recycled oceanic crust in the source of 90–40 Ma basalts in North and Northeast China: Evidence, provenance and significance: Geochim. *Cosmochim. Acta* 143: 49–67.

Xu, Y.G., X.L. Huang, J.L. Ma, Y.B. Wang, Y. Iizuka, J.F. Xu, Q. Wang, and X.Y. Wu. 2004. Crust–mantle interaction during the tectono-thermal reactivation of the North China Craton: constraints from SHRIMP zircon U-Pb chronology and geochemistry of Mesozoic plutons from Western Shandong: Contrib. *Mineralogy and Petrology* 147: 750–767.

Xu, Y.G., X.L. Huang, M.A. Menzies, and G.R. Wang. 2003. Highly magnesian olivines and green-core clinopyroxenes in ultrapotassic lavas from western Yunnan, China: evidence for a complex hybrid origin. *European Journal of Mineralogy* 15: 965–975.

Xu, Y.G., J.L. Ma, F.A. Frey, M.D. Feigenson, and J.F. Liu. 2005. Role of lithosphere–asthenosphere interaction in the genesis of Quaternary alkali and tholeiitic basalts from Datong, western North China Craton: Chem. *Geol.* 224: 247–271.

Xu, Z., Y.F. Zheng, Z.F. Zhao, and B. Gong. 2014. The hydrous properties of subcontinental lithospheric mantle: Constraints from water content and hydrogen isotope composition of phenocrysts from Cenozoic continental basalt in North China. *Geochimica et Cosmochimica Acta* 143: 285–302.

Yan, Q., and X. Shi. 2016. Geochronology of the Laoshan granitic complex in eastern China and its tectonic implications. *Geological Journal* 51: 137–148.

Yan, Q.S., I. Metcalfe, X.F. Shi, P.Y. Zhang, and F.C. Li. 2019. Early Cretaceous granitic rocks from the southern Jiaodong Peninsula, eastern China: implications for lithospheric extension. *International Geology Review* 67: 821–838.

Yang, D.B., W.L. Xu, G.C. Zhao, T.F. Huo, J.P. Shi, and H.T. Yang. 2016a. Tectonic implications of Early Cretaceous low-Mg adakitic rocks generated by partial melting of thickened lower continental crust at the southern margin of the central North China Craton. *Gondwana Research* 38: 220–237.

Yang, J.H., S.L. Chung, M.G. Zhai, and X.H. Zhou. 2004. Geochemical and Sr–Nd–Pb isotopic compositions of mafic dykes from the Jiaodong Peninsula, China: evidence for vein-plus-peridotite melting in the lithospheric mantle. *Lithos* 73: 145–160.

Yang, J.H., J.F. Sun, F. Chen, S.A. Wilde, and F.Y. Wu. 2007a. Sources and petrogenesis of late Triassic dolerite dikes in the Liaodong Peninsula: Implications for post-collisional lithosphere thinning of the eastern North China Craton. *Journal of Petrology* 48: 1973–1997.

Yang, J.H., J.F. Sun, J.H. Zhang, and S.A. Wilde. 2012a. Petrogenesis of Late Triassic intrusive rocks in the northern Liaodong Peninsula related to decratonization of the North China Craton: Zircon U-Pb age and Hf–O isotope evidence. *Lithos* 153: 108–128.

Yang, J.H., F.Y. Wu, S.L. Chung, S.A. Wilde, M. Chu, C.H. Lo, and B. Song. 2005. Petrogenesis of Early Cretaceous intrusions in the Sulu ultrahigh-pressure orogenic belt, east China and their relationship to lithospheric thinning: Chem. *Geology* 222: 200–231.

Yang, J.H., M.F. Zhu, W. Liu, and M.G. Zhai. 2003. Geochemistry and petrogenesis of Guojialing granodiorites from the northwestern Jiaodong Peninsula, eastern China. *Acta Petrologica Sinica* 4 (19): 692–700. (in Chinese with English abstract).

Yang, L.Q., J. Deng, L.N. Guo, Z.L. Wang, X.Z. Li, and J.L. Li. 2016b. Origin and evolution of ore fluid, and gold-deposition processes at the giant Taishang gold deposit, Jiaodong Peninsula, eastern China. *Ore Geology Reviews* 72: 585–602.

Yang, Q.Y., M. Santosh, J.F. Shen, and S.R. Li. 2014. Juvenile vs. recycled crust in NE China: Zircon U-Pb geochronology, Hf isotope and an integrated model for Mesozoic gold mineralization in the Jiaodong Peninsula. *Gondwana Research* 25: 1445–1468.

Yang, X.Z., E. Deloule, Q.K. Xia, Q.C. Fan, and M. Feng. 2008. Water contrast between Precambrian and Phanerozoic continental lower crust in eastern China. *Journal of Geophysical Research: Solid Earth* 113: B8207.

Ye, K., B.L. Cong, and D.N. Ye. 2000. The possible subduction of continental material to depths greater than 200 km. *Nature* 407: 734–736.

Ying, J.F., H.F. Zhang, and Y.J. Tang. 2010. Lower crustal xenoliths from Junan, Shandong province and their bearing on the nature of the lower crust beneath the North China Craton. *Lithos* 119: 363–376.

Zeng, Y.C., J.L. Chen, J.F. Xu, M. Lei, and Q.W. Xiong. 2017. Origin of Miocene Cu-bearing porphyries in the Zhunuo region of the southern Lhasa subterrane: Constraints from geochronology and geochemistry. *Gondwana Research* 41: 51–64.

Zeng, Y.C., J.L. Chen, J.F. Xu, B.D. Wang, and F. Huang. 2016. Sediment melting during subduction initiation: Geochronological and geochemical evidence from the Darutso high-Mg andesites within ophiolite melange, central Tibet. *Geochemistry, Geophysics, Geosystems* 17: 4859–4877.

Zeng, Y.C., J.F. Xu, J.L. Chen, B.D. Wang, Z.Q. Kang, and F. Huang. 2018. Geochronological and geochemical constraints on the origin of the Yunzhug ophiolite in the Shiquanhe–Yunzhug–Namu Tso ophiolite belt, Lhasa Terrane, Tibetan Plateau. *Lithos* 300: 250–260.

Zeng, Y.C., J.F. Xu, F. Huang, M.J. Li, and Q. Chen. 2019. Generation of the 105–100 Ma Dagze volcanic rocks in the north Lhasa Terrane by lower crustal melting at different temperature and depth: Implications for tectonic transition. *Geological Society of America Bulletin,* 10–1130.

Zhai, M.G., and A.G. Bian. 2000. The Neoarchean Super-Continent cohere and Late Paleoproterozoic- Mesoproterozoic disintegration of the North China Craton. *Science in China (Series D)* 30 (S1): 129–137. (in Chinese with English abstract).

Zhang, H. 2012. Destruction of ancient lower crust through magma underplating beneath Jiaodong Peninsula, North China Craton: U-Pb and Hf isotopic evidence from granulite xenoliths. *Gondwana Research* 21: 281–292.

Zhang, H.F., M. Sun, X.H. Zhou, W.M. Fan, M.G. Zhai, and J.F. Yin. 2002. Mesozoic lithosphere destruction beneath the North China Craton:evidence from major-, trace-element and Sr–Nd–Pb isotope studies of Fangcheng basalts. *Contributions to Mineralogy and Petrology* 144: 241–253.

Zhang, H.F., J.F. Ying, G. Shimoda, N.T. Kita, Y. Morishita, J.A. Shao, and Y.J. Tang. 2007. Importance of melt circulation and crust-mantle interaction in the lithospheric evolution beneath the North China Craton: Evidence from Mesozoic basalt-borne clinopyroxene xenocrysts and pyroxenite xenoliths. *Lithos* 96: 67–89.

Zhang, J., Z.F. Zhao, Y.F. Zheng, and M. Dai. 2010. Postcollisional magmatism: Geochemical constraints on the petrogenesis of mesozoic granitoids in the Sulu orogen, China. *Lithos* 119: 512–536.

Zhang, Y.Q., S.W. Dong, and W. Shi. 2003. Cretaceous deformation history of the Tan-Lu fault zone in Shandong province, eastern China. *Tectonophysics* 363: 243–258.

Zhang, Y.Q., J.L. Li, T. Zhang, S.W. Dong, and Y.J. Yin. 2008. Cretaceous to Paleocene tectonosedimentary evolution of the Jiaolai basin and the Contiguous areas of the Shandong Peninsula (North China) and its geodynamic implications. *Acta Geologica Sinica* 82: 1229–1257.

Zhang, Y.Y., C. Yuan, M. Sun, X.P. Long, Y.P. Wang, Y.D. Jiang, and Z.F. Lin. 2017. Arc magmatism associated with steep subduction: Insights from trace element and Sr–Nd–Hf–B isotope systematics. *Journal of Geophysical Research: Solid Earth* 122: 1816–1834.

Zhao, R., Q. Wang, J. Deng, M. Santosh, X. Liu, and H. Cheng. 2018. Late Mesozoic magmatism and sedimentation in the Jiaodong Peninsula: New constraints on lithospheric thinning of the North China Craton. *Lithos* 322: 312–324.

Zhao, R., Q.F. Wang, X.F. Liu, W. Wang, and R.G. Pan. 2016. Architecture of the Sulu crustal suture between the North China Craton and Yangtze Craton: Constraints from Mesozoic granitoids. *Lithos* 266–267: 348–361.

Zhao, Z.F., Y.F. Zheng, R.X. Chen, Q.X. Xia, and Y.B. Wu. 2007b. Element mobility in mafic and felsic ultrahigh-pressure metamorphic rocks during continental collision. *Geochimica et Cosmochimica Acta* 71: 5244–5266.

Zhao, Z.F., Y.F. Zheng, J. Zhang, L.Q. Dai, Q. Li, and X. Liu. 2012. Syn-exhumation magmatism during continental collision: evidence from alkaline intrusives of Triassic age in the Sulu orogen. *Chemical Geology* 328: 70–88.

Zheng, J.P., and F.X. Lu. 1999. Petrologic characteristics of kimberlite-borne mantle xenoliths from the Shandong and Liaoning Peninsula: Paleozoic lithosphere mantle and its heterogeneity. *Acta Geologica Sinica* 15 (1): 65–74.

Zheng, J.P., W.L. Griffin, S.Y. O'Reilly, C.M. Yu, H.F. Zhang, N. Pearson, and M. Zhang. 2007. Mechanism and timing of lithospheric modification and replacement beneath the eastern North China Craton: Peridotitic xenoliths from the 100 Ma Fuxin basalts and a regional synthesis. *Geochimica et Cosmochimica Acta* 71: 5203–5225.

Zheng, J.P., M. Sun, F.X. Lu, and N. Pearson. 2003. Mesozoic lower crustal xenoliths and their significance in lithospheric evolution beneath the Sino-korean Craton. *Tectonophysics* 361: 37–60.

Zheng, J.P., M. Sun, M.F. Zhou, and P. Robinson. 2005. Trace elemental and PGE geochemical constraints of Mesozoic and Cenozoic peridotitic xenoliths on lithospheric evolution of the North China Craton. *Geochimica et Cosmochimica Acta* 69: 3401–3418.

Zheng, Y.F. 2012. Metamorphic chemical geodynamics in continental subduction zones. *Chemical Geology* 328: 5–48.

Zhou, X.H., M. Sun, G.H. Zhang, and S.H. Chen. 2002. Continental crust and lithospheric mantle interaction beneath North China: isotopic evidence from granulite xenoliths in Hannuoba, Sino-Korean craton. *Lithos* 62: 111–124.

Zhou, X.M., and W.X. Li. 2000. Origin of Late Mesozoic igneous rocks in Southeastern China: Implications for lithosphere subduction and underplating of mafic magmas. *Tectonophysics* 326: 269–287.

Zhu, D.C., Q. Wang, P.A. Cawood, Z.D. Zhao, and X.X. Mo. 2017a. Raising the Gangdese Mountains in southern Tibet. *Journal of Geophysical Research: Solid Earth* 122: 214–223.

Zhu, G., Y. Chen, D.X. Jiang, and S.Z. Lin. 2015. Rapid change from compression to extension in the North China Craton during the Early Cretaceous: evidence from the Yunmengshan metamorphic core complex. *Tectonophysics* 656: 91–110.

Zhu, R.X., L. Chen, F.Y. Wu, and J.L. Liu. 2011, Timing, scale and mechanism of the destruction of the North China Craton. *Science China Earth Sciences* 54: 789–797.

Zhu, R.X., H.F. Zhang, G. Zhu, Q.R. Meng, H.R. Fan, J.H. Yang, F.Y. Wu, Z.Y. Zhang, and T.Y. Zheng. 2017b. Craton destruction and related resources. *International Journal of Earth Sciences* 106: 2233–2257.

Zindler, A., and S. Hart. 1986. Chemical geodynamics. *Ann. ReEarth Pl. Science* 14: 493–571.

Zou, D.Y., Y.S. Liu, Z.C. Hu, S. Gao, K.Q. Zong, R. Xu, L.X. Deng, D.T. He, and C.G. Gao. 2014. Pyroxenite and peridotite xenoliths from Hexigten, Inner Mongolia: insights into the Paleo-Asian ocean subduction-related melt/fluid-peridotite interaction. *Geochimica et Cosmochimica Acta* 140: 435–454.

Zou, S.H., D. Xu, T. Deng, Q.Y. Huang, D. Yu, Z. Zhao, and T. Ye. 2019. Geochemical variations of the Late Mesozoic granitoids in the southern margin of North China Craton: A possible link to the tectonic transformation from compression to extension. *Gondwana Research* 75: 118–133.

Chapter 4
Micro-Geochemistry of Pyrite and Gold Mineral in Jiaodong Gold Deposits: Implications for Genesis of Gold Deposit

Abstract The Jiaodong gold province, one of the main gold-producing areas in China, includes disseminated stockwork Jiaojia-type and auriferous quartz vein Linglong-type gold deposits. In this chapter, two types of typical gold deposits, such as Linglong and Jiaojia gold deposits, were carried out to conduct comparative studies aim to reveal genesis of Jiaodong-type gold deposit. Pyrites, the main gold-bearing mineral, has been analyzed by using micro-geochemistry. Micro-geochemistry of gold grains was also analyzed. These works aim to the source of ore-forming fluids and materials.

Keywords Micro-geochemistry · Pyrite · Gold mineral · Ore-forming fluids and materials

4.1 Introduction

The Jiaodong gold province is one of the largest gold-producing areas in China. There are 159 gold mines in the Jiaodong Peninsula producing a combined 30 tons of Au per year (Yang et al. 2016a; Yang and Santosh 2019). There are two main types of gold deposits, Linglong and Jiaojia, in the Jiaodong Peninsula. Linglong-type gold deposits indicate that gold-bearing quartz veins are formed in secondary faults that intrude into Mesozoic granites; one such example includes the Linglong gold deposit. Jiaojia-type gold mineralization occurs mainly disseminated and reticulated, and locate along the first-level regional fracture and fracture-alteration zones; examples include the Sanshandao and Jiaojia gold deposits (Deng et al. 2003; Li et al. 2015; Song et al. 2015). Many previous studies—including those addressing the geological and geochemical characteristics as well as structural characteristics, ages, materials, and fluids—have been conducted on the Jiaodong gold deposits. These studies have estimated that the main ore-forming age of the Jiaodong gold deposit is ~122 Ma (Yang and Zhou 2001; Zhang et al. 2003a, 2020; Li et al. 2006; Hu et al. 2013; Deng et al. 2015b; Wang et al. 2015b; Ma et al. 2017; Bi and Zhao 2017; Cai et al. 2018; Feng et al. 2018). Previous studies have also assumed that ore-forming fluids are part of a medium-temperature, medium-low salinity CO_2-bearing hydrothermal system (Hu et al. 2013; Li et al. 2013; Wang et al. 2015a; Wen et al. 2015; Guo

Y. Liang, *Mineral Micro-Geochemistry Constraints on Petrogenesis and Genesis of Gold Deposit*, https://doi.org/10.1007/978-981-16-1022-6_4

et al. 2017; Cai et al. 2018). Although several studies have evaluated the sources of ore-forming fluids and ore-forming materials (using C-D-O-S-He–Ar isotopes), mineral associations, and mineral geochemistry (Fan et al. 2003; Mao et al. 2008; Hu et al. 2013; Mills et al. 2015a; Wen et al. 2015; Yang et al. 2016a, b; Deng et al. 2015a, 2019; Cai et al. 2018; Feng et al. 2018), the sources of ore-forming fluids and materials and genesis of the Jiaodong gold deposits are still controversial. Especially, whether mantle-derived fluids/materials played major role in the gold mineralization is the major scientific problem and can't be resolved up to now. Thus, the genesis of the Jiaodong gold deposit is still a enigmatic.

The principal gold-bearing mineral in the deposit is pyrite. Therefore, an in-depth analysis of the mineralogical characteristics of the pyrite has theoretical and practical significance for analysing the source of ore-forming materials in this area and developing prospecting predictions. Early studies focused on the shallow position of the deposit in the study area, and studies that address the currently mined deep samples are scant. Pyrite, the most common sulphide, is usually an important symbol of gold mineralisation; its formation process runs through a large-scale gold mineralisation event in Jiaodong. This research is primarily focused on the No. 175 vein of the Jiuqu gold deposit (below 600 m). Systematic analysis of the main mineralogical and structural characteristics of the pyrite in each metallogenic stage and the relationship between pyrite and visible gold is needed to provide a basis for deep prospecting prediction. In addition, analyses of major and trace elements, Sr-Nd isotopes, and in situ S isotopes were conducted on the pyrite of the main gold-bearing mineral assemblages from each mineralization stage of the Linglong and Jiaojia gold deposits. This information may be used to explore the source of ore-forming materials and the process of Au element migration and deposition. The geodynamic setting of mineralisation in the study region is also summarised.

Gold minerals in ores is an important research topic in the study of gold deposits. Scholars in China and abroad have analysed the occurrence state, mineral form, and formation environment of gold through various aspects, such as gold-mineral morphology, mineralogy, and geochemistry and provided strong evidence for the gold deposit formation process (Liu and Zhang 1996; Yu et al. 1997; Pettke et al. 1997; Zeng and Fan 2002; Bonev et al. 2002; Nakagawal et al. 2005; Larizzatti et al. 2008; Cabri et al. 2008; Litvinenko 2009; Liu et al. 2010a). The mineralogical characteristics of gold minerals in gold deposits in the area have also been investigated (Li 1985; Wang et al. 1998; Zhang et al. 2006; Deng et al. 2007; Liu et al. 2010b; Yang et al. 2011). However, the studies concerning gold minerals in the gold deposits in Jiaodong have not considered the geochemistry and gold-mineral microregions. Yang et al. (2013) first studied the geochemical characteristics of individual gold particles in a typical quartz vein-type gold deposit (Linglong gold deposit); a new, effective approach to examine gold minerals in Linglong-type gold deposits and elucidate the genetic mechanism of this type of gold deposits was proposed. Typical altered rock-type gold deposits (Sanshandao, Xinli, and Cangshang) are found in the Sanshandao–Cangshang fault zone in the western Jiaodong Peninsula. In this chapter, the Sanshandao and Xinli super-large gold mines were selected as research objects, and a detailed study of gold-mineral microregions was conducted with the help of scanning electron microscope-energy spectrum analysis (SEM-EDS) and electron probe analysis (EMPA). The study includes a comparison of the geochemical characteristics of the microregions of gold particles in the Linglong gold mine, a detailed description of the mineralogical characteristics of the gold minerals in the altered rock-type gold deposit, and a focus on the geochemical characteristics and indicative significance of the microregions of a single gold mineral. New evidence concerning plausible genetic mechanisms of the altered rock-type gold deposits was revealed.

4.2 Ore Deposit Geology

4.2.1 Linglong Gold Orefield

The Jiaodong Linglong gold orefield extends from Jiuqu in the east, Luoshan in the west, Taishang Village in the south, and Dazhuangzi in the north, with an east to west width of 11 km north to south length of 6 km and area of 66 km^2. From a structural point of view, the mining area is controlled by the Potouqing, Linglong, Yuhuangding, and Jiuqu-Jiangjia faults and can be divided into five deposits: Dongfeng, Dongshan, Xishan, and Jiuqu. The lingering gold ore field is a super-large vein-type gold orefield. The Linglong gold deposit is structurally controlled through the Potouqing fault, with a NEE- to NNE-trend, and the Linglong fault, with an NNE-trend. Approximately 200 gold veins are present in the ore field. The exposed strata in the mining area primarily include the metamorphic basement of the Archean Jiaodong Group and Quaternary sediments. The Archean Jiaodong Group primarily includes Precambrian metamorphic amphibolite, metagranulite, and granulite, which are generally exposed in mining areas, primarily river sediments, lake sediments, slope deposits, and floods by accumulation. The magmatic activity in the mining area is relatively strong, primarily showing Early Cretaceous Guojialing and Late Jurassic Linglong granites. Linglong granite is primarily distributed in the south and north of the Qixia compound anticline with an elliptical, round rock mass that has an irregular ring rock base and is distributed in bands and accompanied by certain alterations, such as potash and silicification. The Guojialing granites are primarily distributed in the Qixia compound anticline, and the rock masses are generally NEE-oriented rock strains (Fig. 7.18). The dykes in the deposit, showing a NNE-trend, consist of quartz diorite, diorite, and lamprophyre (Fig. 4.1a).

Orebodies spread at a NNE-trend and have thicknesses varying from 0.8 to 2.0 m. Most of the orebodies are produced in the form of veins with a few lenses (Fig. 4.1b). The ore veins near the Potouqing fault show a NW trend in the deep part and an EW-trend in the shallow part. The No. 108 vein is distributed with an elevation of approximately −400 m. However, the 52 and 47 veins, which are close to the Potoxin fault, are distributed at an elevation of approximately −1100 m. The elevation of the orebody in the Linglong gold deposit has a trend of gradually decreasing from west to east and from north to south. The closer a deposit is to the Potouqing fault, the lower

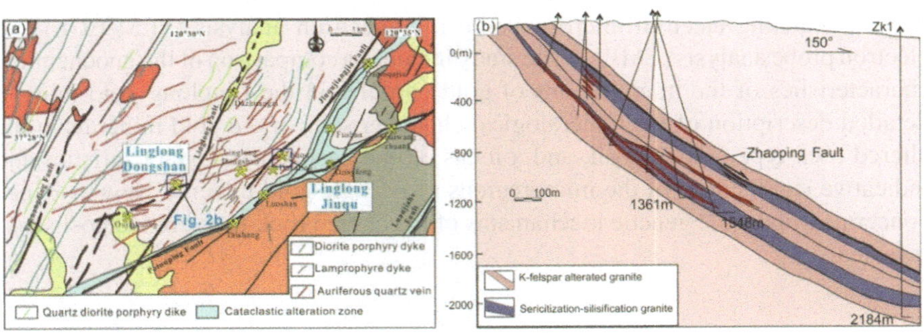

Fig. 4.1 **a** Simplified geological map of the Linglong goldfield (after Yang et al. 2016b), **b** Geological cross-sections along line 72 in the Linglong goldfield (following Shen et al. 2016)

the elevation of the ore bodies (Fig. 4.1b). Ore minerals are composed of main metal minerals (e.g., gold-silver ore and pyrite with minor arsenopyrite, galena, and chalcopyrite) and non-metallic minerals (e.g., calcite, feldspar, chlorite, sericite, quartz, etc.) (Fig. 4.2a–d). The main types of wall-rock alterations include sericitisation, K-feldspar alteration, pyritisation, silicification, beresitisation, and carbonatisation, as well as minor chloritisation and kaolinitisation (Fig. 4.2e–h). Four endogenetic stages were identified from the mineralisation process: (1) Milky white quartz stage: This stage was characterised by a large-scale milky white quartz vein in the fissuring and euhedral pyrites. (2) Quartz-pyrite stage: In this stage, pyrite appeared at the edge or centre of the smoky quartz veins, and abundant electrum and native Au were observed in the fissures. (3) Polymetallic sulfide stage: The development of polymetallic sulfide and high gold content were the main features of this stage. The mineral combination was pyrrhotite, sphalerite, and galena. (4) Quartz–carbonate stage: The presence of calcite was the main feature of this stage. Silver-gold ore and quartz were found.

Fig. 4.2 **a–d** Photomicrographs of gold ore in the Linglong gold deposit, showing important ore minerals; **e–h** Photographs of alteration types in the Linglong gold deposit: **e** K-feldspar alteration, **f** sericitisation, **g** silicification, and **h** carbonatisation

4.2.2 *Jiaojia Gold Deposit*

The Jiaojia gold deposit is located in the middle section of the Jiaojia Fault and has reserves of >300 t (Fig. 4.3a). The ore bodies also are hosted by the Late Jurassic Linglong granite. Lamprophyre and diorite porphyry dykes were also emplaced into the Jiaojia gold deposit (Fig. 4.3b).

Five main orebodies (I, II, III, IV, and V) were identified in the Jiaojia gold deposit. No. Orebody I was close to the major fault and was controlled by phyllic altered cataclastic rocks. The orebody was layer-like, lenticular, and vein-like, running parallel with the major fault and at a NE 10°–50° orientation, a NW trend, and a dip angle of 30° in the shallow area; however, this angle decreased to 16° from −1000 m in elevation. Orebody II is located in the lower branch of the No. I orebody, and was mostly produced from phyllic altered cataclastic rocks. Its spread was parallel with the major fault, and the occurrence of the orebody was discontinuous and showed a lenticular and spindle shape, transitioning from a veinlet-disseminated orebody to a vein orebody. Orebody No. III was distributed in the beresitisated granite, and spread with a similar trend on the major fault surface; it also showed a SE tendency and a dip of 60°–90°. Orebody No. IV was distributed in the K-feldspar alteration and silicification granite belt and was lenticular and vein-shaped. Its occurrence was basically consistent with the major fault, with a tendency of 292°–316° and a dip angle of 16°–30°. Orebody No. V was distributed in the phyllic altered cataclastic rocks, and was located in the upper part of the major fault with a lenticular and veined shape (Fig. 4.3b). The ore minerals were composed of primary metal minerals (e.g., gold-silver ore, pyrite, siderite, chalcopyrite, galena, sphalerite, and specularite, with minor pyrrhotite and arsenopyrite) and non-metallic minerals (e.g., sericite, quartz, calcite, and feldspar, with minor chlorite and apatite) (Fig. 4.4a–d). The pre-mineral alteration assemblage consists of K-feldspar alteration and albitisation, while the gold mineralization was accompanied by phyllic alteration and

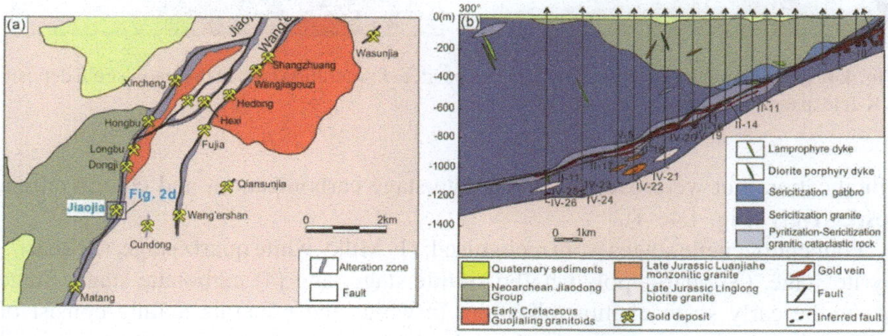

Fig. 4.3 **a** Sketched geologic map of the Jiaojia gold field (Modified from Yang et al. 2016a), and **b** geological cross-sections along line 112 in the Jiaojia gold field (after Chen 2017)

Fig. 4.4 **a–d** Ore minerals in the Jiaojia gold deposit; **e–f** Alteration type in the Jiaojia gold deposit: **e** K-feldspar alteration and **f** sericitisation

silicification that were overprinted by late-stage carbonization, and chloritization, respectively (Fig. 4.4e–f).

Four endogenetic stages were recognised: (1) Milky white quartz stage, (2) quartz-pyrite stage, (3) quartz–polymetallic sulfide stage, and (4) carbonate stage. Stage (1) is the early stage of mineralisation, in which the minerals mainly consist of milky white veined quartz and euhedral pyrite. Stage (2) was a primary stage of mineralisation. Smoky grey quartz and pyrite were developed in the beresite cracks or altered cataclastic rocks and were generally disseminated with veined shapes.

Stage (3) was also a primary stage of mineralisation. Quartz and polymetallic sulfide (e.g., galena, pyrite, and sphalerite) were developed in the phyllic altered cataclastic rocks and were generally disseminated with veined shapes. Stage (4), the late stage of mineralisation, filled in the early fissures. The minerals mainly consisted of dolomite, calcite, and a small amount of generally veiny and blocky quartz and pyrite.

4.2.3 Sanshandao-Xinli Deposit

The "Sanshandao–Cangshang" fault zone is located in Sanshandao Town, Laizhou City, northwest Jiaodong Peninsula and runs towards the NNE–NE. The fault zone controls the three large deposits of Sanshandao, Xinli, and Cangshang (Fig. 4.5a, b). Xinli, Sanshandao, and Cangshang gold deposits are altered rock-type gold deposit.

The Xinli gold deposit has four veins (I–IV veins), with a proven metal content of 37 t and average gold grade of 3.29 g/t, of which vein I is the principal vein,

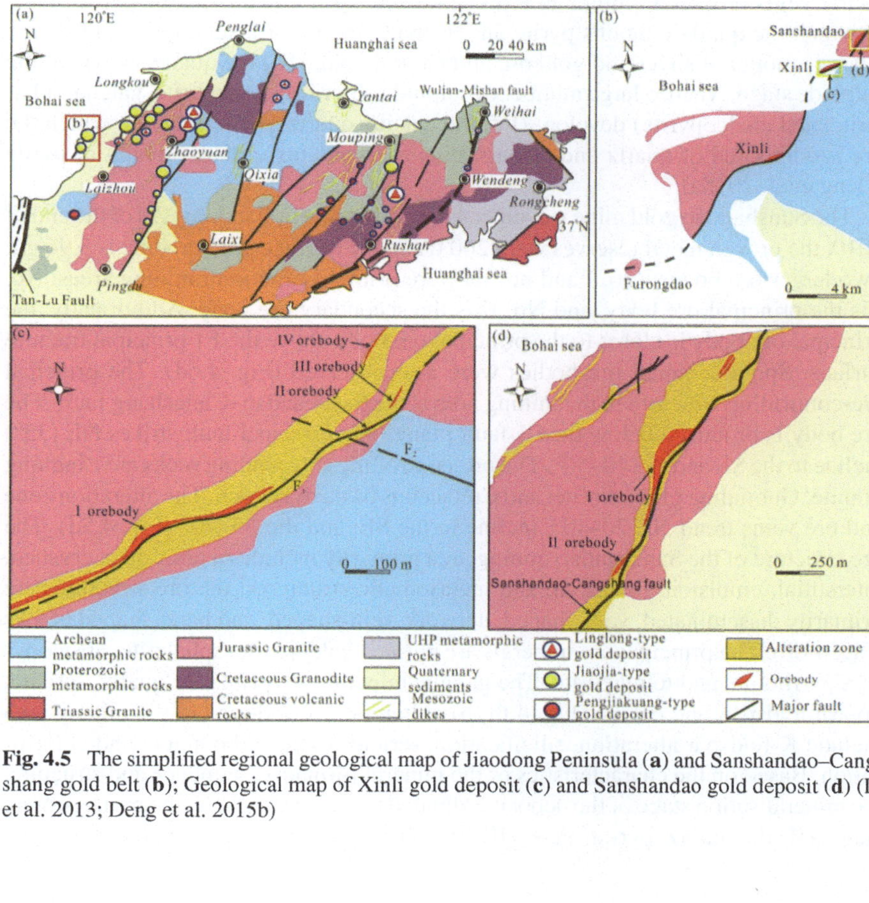

Fig. 4.5 The simplified regional geological map of Jiaodong Peninsula (**a**) and Sanshandao–Cangshang gold belt (**b**); Geological map of Xinli gold deposit (**c**) and Sanshandao gold deposit (**d**) (Li et al. 2013; Deng et al. 2015b)

accounting for 93% of the total ore reserves. The ore body and alteration zone strike in the NE direction (20°–40°), incline towards the SE, and dip 35°–70° (Fig. 4.5c). The ore-controlling structure in the mining area is the primarily NNE–NE-trending Sanshandao–Cangshang fault and its secondary faults. In the Xinli mining area, the principal fault F1 is approximately 1300 m long, 70–185 m wide and has a SE incline of 40°–50° (Fig. 4.5c). The ore-controlling surrounding rocks are primarily Linglong granite and Guojialing granodiorite; a certain scale of basic dyke rock is associated with the ore body. The ore structure of the Xinli gold-mining area primarily includes grain, interstitial, crushed, and metasomatic residual structures (Fig. 4.6a,). The principal structures are disseminated, vein, mass, and network vein. The principal ore minerals are pyrite, chalcopyrite, sphalerite, galena, pyrrhotite, and tetrahedrite; and the gangue minerals are primarily quartz, sericite, calcite, potash feldspar, and rutile (Fig. 4.6 e, g). The alteration types of ore deposits include K-feldspar alteration, silicification, pyrite sericification, carbonation, and chloriteisation. Based on the mineral symbiosis characteristics, mineral formation sequence, and ore structure, the mineralisation of the deposit is divided into four principal metallogenic stages: (I) the quartz-sericite-pyrite stage, in which euhedral pyrite ore is filled in the milky white quartz veins or quartz clumps in a disseminated state; (II) the quartz–pyrite stage, in which the quartz contains pyrite and a small amount of arsenopyrite and pyrite, a large amount of silver and gold ore and natural gold; (III) the quartz–polymetallic sulphide stage, where a large number of sulphides (pyrite, arsenopyrite, galena, sphalerite, and chalcopyrite) develop; (IV) the calcite–quartz–pyrite stage, in which the ore is composed of quartz and calcite veins and contains a small amount of pyrite (Deng et al. 2015a).

The Sanshandao gold mine is a super-large gold mine in Jiaodong. As of the end of 2010, the proven metal reserves were 200 t, and the average gold product is 3.96 g/t. Two larger ore bodies (no. I and no. II) were primarily found in mining areas. No. I is the principal ore body, and No. II is the secondary ore body. Additionally, the principal ore body is closer to the principal ore body above the F1 principal fracture surface. Sporadic small ore bodies were also observed (Fig. 4.6d). The principal ore-controlling structure in the mining area is the Sanshandao–Cangshang fault. The ore body is produced below the F1 fault plane. The principal fault strikes NE (35°) incline to the SE and dip 40–50°. The ore-controlling surrounding rocks are Linglong granite, Guojialing granodiorite, and Cretaceous basic dyke rock. The alteration zone and ore veins trend NE 20–40°, incline to the SE, and dip 35–70° (Fig. 4.5d). The ore structure of the Sanshandao mining area primarily includes a grain-like, crushed, interstitial, emulsion, inclusion, and metasomatic structures; the ore structures are primarily disseminated, vein-shaped, network-vein-shaped, and band-shaped blocks (Fig. 4.6b). The principal ore minerals are pyrite, chalcopyrite, sphalerite, arsenopyrite, pyrrhotite, and tetrahedrite. The gangue minerals are primarily quartz, sericite, potash feldspar, and calcite (Fig. 4.6f, h). Alterations developed in the mining area include K-feldspar alteration, silicification, sericitisation, carbonation, and chloritisation. Based on the characteristics of the mineral formation sequence and structure, the mineralisation stage of the deposit is divided into four stages: (I) the pyrite sericite stage, (II) the quartz–pyrite stage, (III) the quartz–polymetallic sulphide stage, and

Fig. 4.6 Photograph and Photomicrographs of ore in Xinli and Sanshandao gold deposit. Photograph of ore in Xinli gold deposit (**a**, **c**) and Sanshandao gold deposit (**b**, **d**); Photomicrographs of gold grains in ore from Sanshandao gold deposit (**f**, **h**) and Xinli gold deposit (**e**, **g**). Py: Pyrite, Gn: Galena, Sp: Sphalerite, Ccp: Chalcopyrite

(IV) the quartz–calcite stage. The samples studied in this paper were collected from the No. 35 penetrating vein (X0635-15) in the middle section of the Xinli gold mine −240 m (X0635-15) and the No. 132 penetrating vein (S09132-5) in the middle section of the Sanshandao gold mine −420 m.

4.3 The Micro-Geochemical Characteristics of Gold from "Jiaojia-Type" Gold Deposit: Implications for Sources of Ore-Forming Minerals and Fluids of Sanshano-Xinli Gold Deposit

4.3.1 Mineralogical Characteristics

Optical microscopy observations show that the principal mineral combinations in gold-bearing mineral samples include pyrite, chalcopyrite, sphalerite, and galena, which belong to the quartz–polymetallic sulphide stage (stage III); pyrite is the principal gold-bearing mineral (Fig. 4.6). Gold primarily exists in gold-bearing minerals in the form of single particles, with an obvious golden yellow reflection colour, high reflectivity, homogeneity, and low hardness and are completely different from other metal sulphides (Fig. 4.6c–f). To more clearly observe the morphology of gold minerals, a high-power scanning electron microscope (SEM) (from 9000× to 20000×) were used. In the BSE photos, gold minerals primarily exist in the form of flakes, ellipses, short columns, and plates, with a length of approximately 15–25 μm (Fig. 4.7a–d). Three principal occurrence states of gold minerals exist in sulphides: interstitial gold, fractured gold, and coated gold. Interstitial gold—pyrite interstitial and short columnar and plate-like gold minerals—developed between the interstitials of pyrite and polymetallic sulphide and presented a symbiotic relationship (Fig. 4.7d, e); fissure gold, the principal gold mineral, exists in the cracks formed by the fragmentation of pyrite or polymetallic sulphides in flakes or an elliptical form (Fig. 4.7c); the gold mineral is wrapped in the pyrite in an elliptical form (Fig. 4.7f).

The BSE map reveals that the surfaces of gold minerals do not have uniform chromaticity but feature brightly coloured and dark grey areas. The SEM secondary electron image revealed that the dark grey areas in the gold minerals are the pits on the mineral surface and that the bright areas are the shiny gold surfaces. SEM high-energy imaging has revealed that the original natural gold particles have a smooth and regular granular structure; however, upon precipitation of hydrothermal fluid, the surface of the particles corrodes and decomposes, thus showing grooves or pits (Nakagawal et al. 2005). Similarly, the formation of the probe sheet could create pits when squeezed by an external force. In the secondary electron image, the lighter pit and lighter coloured area may be artificially caused by the grinding of the film; however, the darker and deeper areas may be the pits of the gold mineral itself. Therefore, this article studies

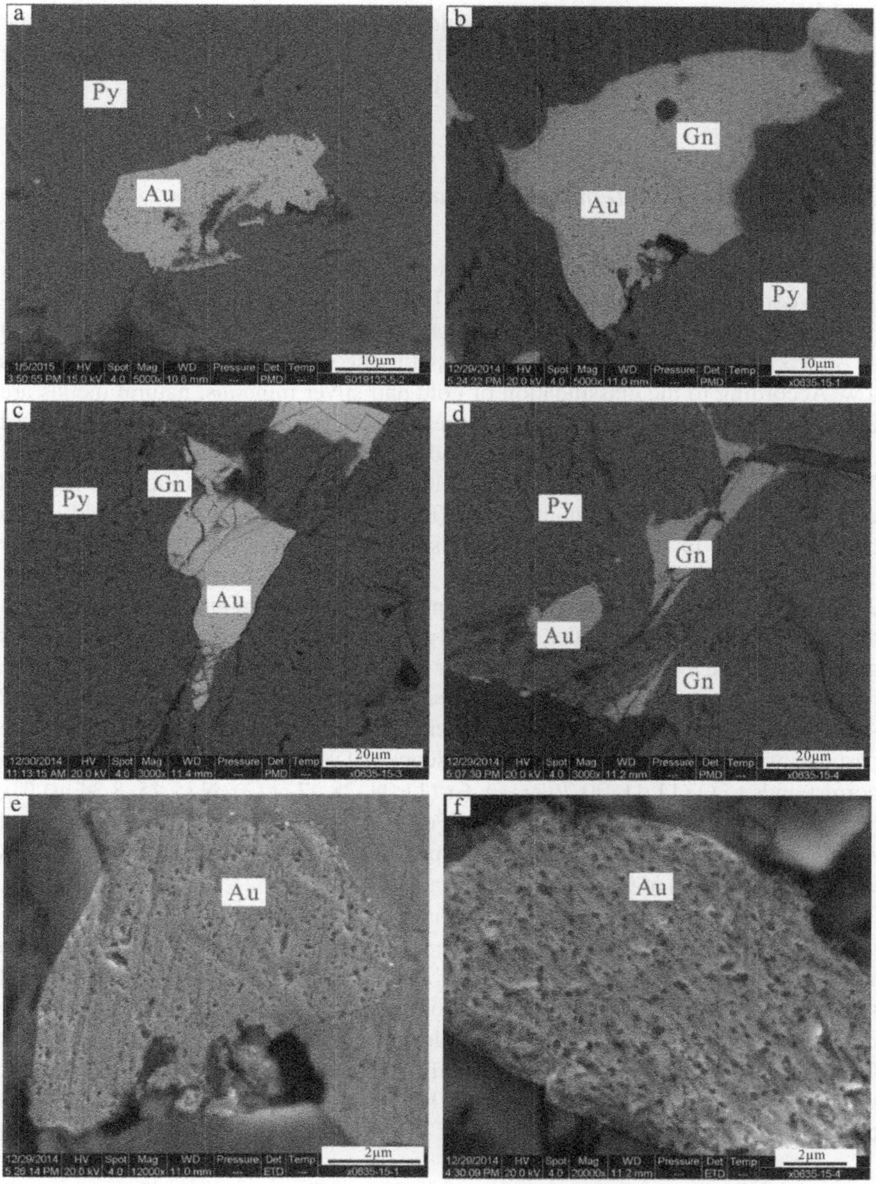

Fig. 4.7 Back-scattered electron images of Au grains (**a**–**d**); Secondary electron image of Au grains (**e**, **f**) in the Sanshandao and Xinli gold deposits

the dark grey areas on the surface of gold minerals, which may be caused by the reflection of the surface pits or by the filling of the pits with other substances. The chemical composition is further explored below (Fig. 4.7e, f).

4.3.2 Micro-Geochemical Characteristics

The electron probe (EMPA) analysis results are expressed as mass percentage (wt%). The Au content in gold minerals is 63.34–72.54 wt%, with a high Ag content (27.39–36.34 wt%) (Table 4.1). According to the calculation of natural gold fineness [Au/(Au + Ag)] × 1000, the fineness of natural gold in the study area is 640–726, which is low. According to the classification of gold and silver minerals, natural metals are found in a series of silver and gold mines (Liu et al. 2010a). Therefore, the gold minerals in this study are from silver and gold mines. The Au and Ag elemental X-ray scanning images of the gold particles revealed a high content of Ag and of gold-bearing minerals containing Ag (Fig. 4.8). The series of silver-gold mines generally refers to primary gold minerals, which have not been affected by later environments and can represent the initial generation environment (Nakagawal et al. 2005). When scanning electron microscopy-energy spectrum analysis (SEM-EDS), the geochemical characteristics of the silver-gold ore micro-area are expressed as atomic percentage (at.%). The high-power BSE electron picture reveals two obvious bright and dark grey areas on the surface of the silver-gold ore particles. In this study, brightly coloured and dark grey microregions on the surface of silver-gold ore particles were detected by energy spectroscopy (EDS) (Fig. 4.9). The dark grey area corresponds to O (11.41–20.26 at.%), Cr (11.08–28.47 at.%), N (5.45–8.34 at) %), W (0.22–0.74 at.%), P (4.29–8.43 at.%), Fe (1.58–5.27 at.%), and a small amount of Al and S; the brightly coloured area corresponds to N (6.12–9.79 at.%), O (1.89–6.18 at.%), W (0.42–0.84 at.%), P (6.32–10.38 at.%), and Fe (1.62–3.37 at.%) (Table 4.2). The trace element composition revealed that the biggest difference between the dark grey and brightly colour areas is that the dark grey areas are rich in Cr, O, and a small amount of Al and S. The dark grey areas are observed as pits on the mineral surface under the SEM secondary electron image (Fig. 4.7e, f). According to the aforementioned sample processing procedure, the final polishing of the sample uses Cr_2O_3; thus, we cannot discard the possibility that the Cr and O contained in the dark grey area are impurities. In addition, in the bright colour area, no Cr was detected, and the content of O was significantly reduced. Because the bright colour area is a natural

Table 4.1 Electronic microprobe analysis(wt%) of gold grain from sample X0635-15 and S09132-5 in the Sanshandao and Xinli gold deposits

Sample	Ag	Au	Total
X0635-15-3-1	36.04	64.15	100.19
X0635-15-4-1	35.73	64.27	100
X0635-15-1-1	36.34	63.34	99.69
S09132-5-2-1	27.39	72.54	99.93

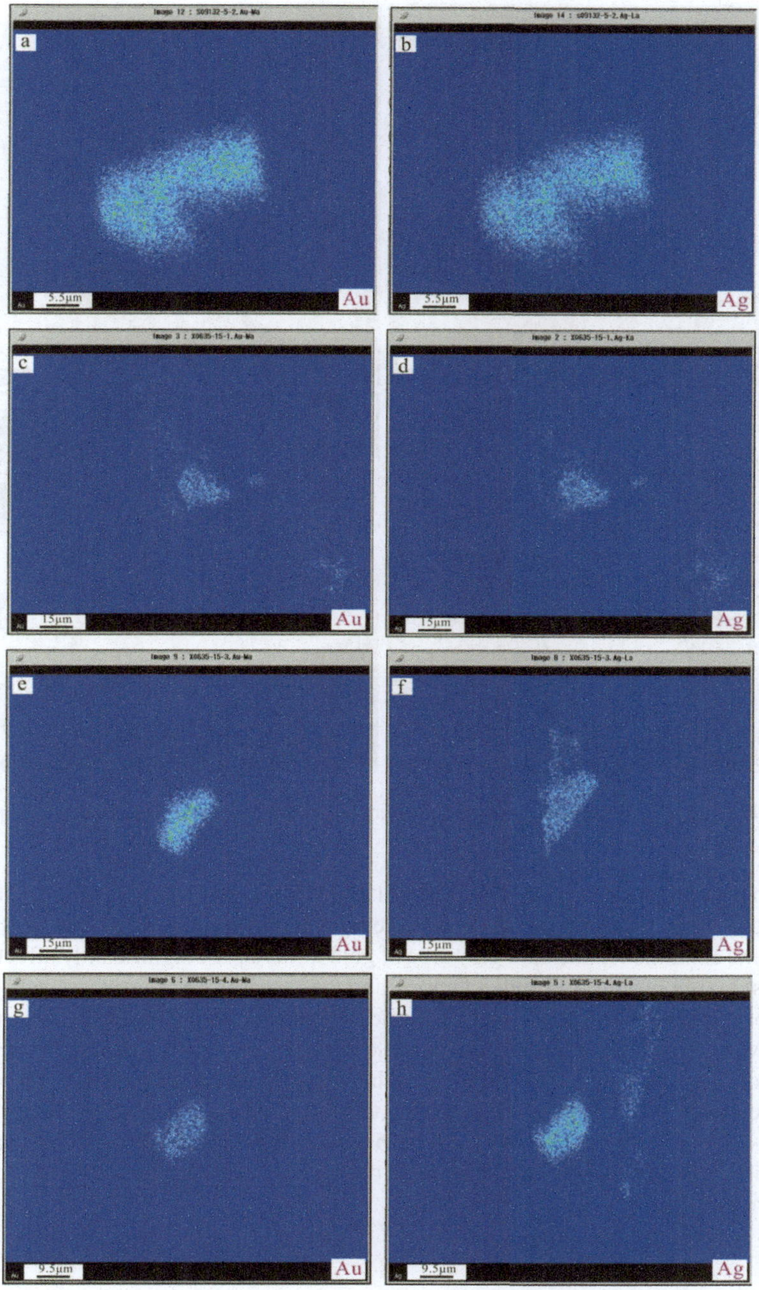

Fig. 4.8 X-ray scanning images of Au and Ag in the Sanshandao and Xinli gold deposits

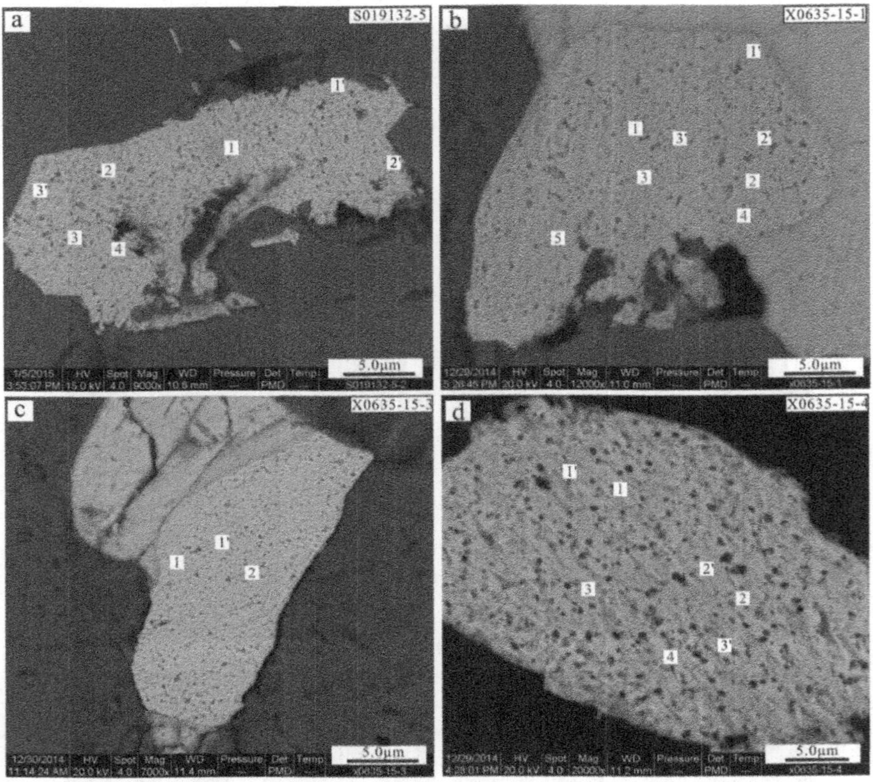

Fig. 4.9 Back-scattered electron images of Au grains in the Sanshandao and Xinli gold deposits

gold-polished surface, there is no residual Cr_2O_3, further supporting the possibility Cr and O high content in the dark grey area derived from impurities. Therefore, the elements contained in the dark grey area of gold minerals are significantly affected by the residual impurities of human factors, and the research error is excessively large. This paper primarily studies the bright microregions of gold particles and compares them with the bright microregions of gold particles in the Linglong gold deposit. The gold particles in the study area were brightly coloured micro-areas. Au and Ag are the principal elements. The principal trace elements are N (6.12–9.79 at.%), O (1.89–6.18 at.%), W (0.42–0.84 at.%), P (6.32–10.38 at.%), and Fe (1.62–3.37 at.%); there is no obvious content change from the inner micro-area of the particle to the edge micro-area.

Table 4.2 SEM–EDS analysis of gold and silver grains from in the Sanshandao and Xinli gold deposits in Jiaodong peninsula. (Data are from Liang et al. 2014)

Sample	Area	No.	B	N	O	Al	Si	P	S	Fe	Cr	Zn	Nb	Ag	W	Au	Data sources
X0635-15-1	Dark	1'		7.32	12.29			6.51			11.08			27.98	0.73	34.1	Liang et al. (2014)
	Dark	2'		7.17	16.34			6.22			15.93			24.25	0.6	29.49	Liang et al. (2014)
	Dark	3'		6.64	15.84			5.29			14.2			25.97	0.22	31.83	Liang et al. (2014)
	Bright	1		8.04	3.26			8.01		1.8				36.8		42.09	Liang et al. (2014)
	Bright	2		8	2.39			7.19		2.17		1.74		37.86		40.65	Liang et al. (2014)
	Bright	3		9.11	2.63			7.38						36.76		44.12	Liang et al. (2014)
	Bright	4		7.3	2.82			8.57						37.96		43.35	Liang et al. (2014)
	Bright	5		8.73	2.86			7.52						38		42.89	Liang et al. (2014)
X0635-15-3	Dark	1'			16.37			6.08			15.74			27.78	0.42	33.6	Liang et al. (2014)
	Bright	1		7.34	3.05			8.18		1.62				37.28	0.84	41.7	Liang et al. (2014)
	Bright	2		7.22	1.89			10.37		1.97				35.21	0.6	42.74	Liang et al. (2014)
X0635-15-4	Dark	1'		6.43	17.37			5.45		1.82	24.13			18.08	0.68	26.03	Liang et al. (2014)
	Dark	2'		6.98	15.02			6.67		1.58	15.93			22.64	0.74	30.43	Liang et al. (2014)
	Dark	3'		6.45	12.89			6.49		2.09	16.26			24.3	0.27	31.26	Liang et al. (2014)
	Bright	1		8.03	4.26			8.5		3.37				35.61		40.22	Liang et al. (2014)
	Bright	2		8.98	5.46			8.28		2.5				33.81		40.96	Liang et al. (2014)
	Bright	3		8.56	6.18			7.65		2.79				33.52	0.67	40.62	Liang et al. (2014)
	Bright	4		7.37	4.51			7.82		3.16				34.59	0.73	41.82	Liang et al. (2014)
S09132-5-2	Dark	1'			13.55			5.87			18.61			17.66		44.31	Liang et al. (2014)
	Dark	2'		5.45	20.26	1.28		4.29	2.45	5.27	28.47			10.69			Liang et al. (2014)
	Dark	3'		8.34	11.41			8.43		3.17	13.22			17.11	0.67	37.64	Liang et al. (2014)

(continued)

Table 4.2 (continued)

Sample	Area	No.	B	N	O	Al	Si	P	S	Fe	Cr	Zn	Nb	Ag	W	Au	Data sources
	Bright	1		6.12				6.32						20.61	0.42	66.53	Liang et al. (2014)
	Bright	2		8.67	3.1			9.52		2.31				25.63	0.66	50.09	Liang et al. (2014)
	Bright	3		9.79	2.95			10.02		2.8				26.3	0.54	47.59	Liang et al. (2014)
	Bright	4		8.58	3.38			10.38		2.91				25.48	0.78	48.5	Liang et al. (2014)
Linglong gold	Bright	0’			23.58									15.05		61.37	Yang et al. (2013)
	Bright	1’		20.09	29.24					5.2	2.54			7.82		35.11	Yang et al. (2013)
	Bright	2’		16.48	15.81					4.63				12.74		50.34	Yang et al. (2013)
	Gray	3’			70.37								1.41	5.85		22.36	Yang et al. (2013)
	Gray	4’			69.93								1.67	5.85		22.56	Yang et al. (2013)
	Gray	1			69.18								1.48	5.93		23.42	Yang et al. (2013)
	Gray	2			75.83								1.3	4.29		18.58	Yang et al. (2013)
	Gray	3		12.77	49.14					3.23				6.95		27.91	Yang et al. (2013)
	Dark	4		8.83	48.03					1.53	22.6			2.79		16.22	Yang et al. (2013)
	Dark	5	91.33	2.18	1.16			0.59		0.56	0.15			0.54		3.48	Yang et al. (2013)
	Dark	6		18.89	11.39		29.63	3.54		0.68				5.6		30.95	Yang et al. (2013)
Linglong silver a	Bright	0							1.22					65.85			Yang et al. (2013)
	Bright	1												54.15		12.45	Yang et al. (2013)
	Gray	2			67.41				0.49		26.23			3.98			Yang et al. (2013)
	Gray	4			56.23	1.49	5.09		2.53					23.12			Yang et al. (2013)
	Gray	6			24.68		1.35		2.61					47.2			Yang et al. (2013)
	Dark	7	95.07		3.34	0.48	0.27				0.3			0.37			Yang et al. (2013)

(continued)

Table 4.2 (continued)

Sample	Area	No.	B	N	O	Al	Si	P	S	Fe	Cr	Zn	Nb	Ag	W	Au	Data sources
	Dark	8	97.85		0.39									0.93		0.25	Yang et al. (2013)
Linglong silver b	Gray	0			57.76		0.78		0.94		16.84			15.75			Yang et al. (2013)
	Gray	1			40.46	1.25	2.37		1.06		2.8			32.63			Yang et al. (2013)
	Gray	2			37.64				2.25		4.58			36.93			Yang et al. (2013)
Linglong silver c	Gray	1			67.95				1.91	1.13	24.22			3.56			Yang et al. (2013)
	Gray	2			11.32				3.78	4.67	1.4			53.93			Yang et al. (2013)

4.3.3 Geological Significance of Elements in Silver-Gold Minerals

Yang et al. (2013) found that the surfaces of gold minerals in the Linglong gold deposit have different colours, which are divided into bright areas, light grey areas, and dark grey areas. For the gold particles, bright areas represent gold and contain a small amount of silver; light grey areas are rich in O and contain a small amount of Nb; dark grey areas are rich in B (up to 90%) and Cr. Silver particles are similar to those of gold particles. The light grey area is enriched with O, and the dark grey area is enriched with B. However, in this study, the single-particle surface of the silver-gold mine has only bright and dark grey areas, and the dark grey areas are affected by impurities. This paper studies the comparison of elements in the bright-coloured area of the silver-gold particles and the bright-coloured area of the gold particles in the Linglong gold deposit. This study did not detect the presence of Nb and B, O content was relatively low, and Cr content was similar. Au, Ag, N, O, and Fe remained unchanged; however, the contents of N, O, and Fe contained in the gold particles in the study area are much lower than those in the Linglong gold deposit. W was detected in the gold particle microregions of the study area; however, this element was not found in the gold particle microregions of the Linglong gold deposit.

(1) O

The dark grey area of the gold minerals in the study area was affected by impurities (Cr_2O_3) and had abnormally high Cr and O. However, no elemental Cr was detected in the bright colour area, and the content of O was substantially reduced. Therefore, elemental O contained in the bright colour area is not affected by O in the impurity; the bright colour is because of the O contained in the gold mineral itself. The precipitation of Au is completed as the fluid environment changes. Environmental changes lead to changes in the physical parameters of the ore-forming fluid, which leads to Au precipitation. During the ore-forming process, the ore-forming fluid transitions from an oxidising to reducing environment as the fluid parameters change (Deng et al. 2005; Yang et al. 2013). The low O atom content (at.%) of gold particles in the micro-area do not directly define the source but can be indicative of the environment. Comparing the O content in the gold-mineral microregions of the Linglong gold deposit revealed that the gold-mineral microregions of the altered rock-type and quartz vein-type gold deposits have lower O, indicating the occurrence of gold precipitation in a reducing environment; the percentage of O atoms in the gold minerals of the Linglong gold deposit is relatively lower than that in the gold-mining area of the study area. The gold-forming environment in the study area is more reducible. Additionally, the content of O and Fe in the micro-area of Linglong gold mine is higher than that in the study area; thus, the gold precipitation environment of the Linglong gold mine is less reducible. Combined with the analysis of the structural environment of the gold deposit, the Linglong-type (i.e. quartz vein-type) gold deposit was formed in the shallow brittle fracture zone, and the Jiaojia type (i.e. altered rock type) was formed in the deep transition period from ductile to brittle deformation (Deng et al. 2005). Therefore, a comparison of the O content in gold-mineral microzones indicates that the gold in the Jiaodong gold deposit was deposited in a partial reduction environment and that the reduction of the gold precipitation environment in the altered rock-type gold deposit is slightly higher than that in the quartz vein-type gold deposit.

(2) Ni

Nitrogen is the most distributed element in nature after oxygen and is connected to gold deposits through ore-forming fluids. Ore-forming fluids can be divided into surface and internal fluids. Surface fluids primarily include atmospheric precipitation, seawater, and surface brine. Internal fluids refer to fluids produced in the lithosphere and upper mantle sources (Lu et al. 2004; Deng et al. 2005).

Internal ore-forming fluids contain many N_2 sources. The ore-forming fluid reacts with the surrounding rock, and the fluid directly captures N_2 from the surrounding rock (Mernagh 2001). In the process of contact metamorphism when magmatic fluid is added to the surrounding rock, the NH_4 + silicate minerals in the surrounding rock can produce N_2 (Fu et al. 2014). During the mineralisation process in the study area, strong and massive water-rock reactions occur when fluid is added to the surrounding rock, which is an important stage and mechanism of mineralisation (Jiang et al. 2011; Li et al. 2013; Wen et al. 2015). If N originates from the surrounding rock, the reaction should be large-scale and can be detected in fluid inclusions; however, N_2 has not been found in existing fluid inclusion studies in the study area. Therefore the gold-mineral microzone N is unlikely to have originated from the surrounding rock. Additionally, N in the deep crust replaces K in the form of NH_4 + and exists in minerals or in the form of N_2 as a free fluid phase (Andersen et al. 1993). According to McDonough (2003), the N abundance in the primitive mantle is low—approximately 2 μg/g. Moreover, the principal composition of fluid in mantle xenolith inclusions is pure CO_2, and N_2, CO, SO_2, and inert gases that are only measured accidentally (Andersen et al. 1990; Huraiova et al. 1991; Zhang et al. 2004). Therefore, N in deep crust or mantle fluid should exist as N_2, which is not present in fluid inclusions in the study area, perhaps because of excessive scarcity. Another possibility is that N_2 is separated in the form of volatile matter during the ascent, as changes in temperature and pressure lead to changes in the physical and chemical properties of the ore-forming fluid. Therefore, N detected in the microzone of the silver-gold ore particles is unlikely to have originated from the deep fluid.

N also occurs in large amounts in surface fluids. Studies on the fluid inclusions in the Xinli and Sanshandao gold deposits reveal that the ore-forming fluids are primarily fluids with medium-low temperature and low salinity; density gradually decreases from the early to later stages of mineralisation. The relationship between seawater and brine is not significant and is likely controlled by the addition of low-density atmospheric precipitation. Atmospheric precipitation dissolves CO_2, O_2, N_2, and other gases in the atmosphere when it falls, and river and lake water also come from atmospheric precipitation; N elements exist in the form of NO_3–(Lu et al.

2004). Therefore, elemental N in the gold particles may be N_2 or NO_3-from atmospheric precipitation added to the ore-forming fluid in the later stage of mineralisation, thereby being present in gold minerals in trace amounts. This is consistent with the discovery of D-O isotope studies in this area that ore-forming fluids mixed with atmospheric precipitation in the later period (Fan et al. 2003; Li et al. 2013; Deng et al. 2015b).

Therefore, elemental N in the silver-gold mine micro-area reflects the mixture of the ore-forming fluid with atmospheric precipitation in the later period. Yang et al. (2013) found that elemental N exists in the gold-mineral microzones of the Linglong gold deposit, which is higher than in the study area and can reflect the relatively high degree of mixing of the ore-forming fluids of the Linglong gold deposit with atmospheric precipitation.

(3) P

Experimental studies have shown that P can exist in the form of ion coordination in octahedral gold particles (Southam and Beveridge 1996). In geochemistry, P is a high-field element with relatively stable chemical properties. P is not compatible with mantle minerals during partial melting in the mantle but is enriched in the melt and is a compatible element (Roilison 1993). P and its hydrides have high volatility and are also important members and characteristic marker elements of mantle fluids that have a high content of molten inclusions in mantle xenoliths (Du 1996). In basalt—especially alkaline basalt—the P_2O_5 content is higher. For example, the inclusions of Wudalianchi volcanic rock magma (bishonite) are rich in S, P, F, and Cl volatiles (Xia et al. 1990). Therefore, P can exist in mantle fluid in the form of volatile matter. P_2O_5 in the alkaline basalt of Dongfangcheng in the existing mountain near the study area is approximately 0.74–0.96, with an average of 0.84% (wt%) (Zhang et al. 2002; Guo et al. 2013a, b), which is much higher than that of P_2O_5 in the Linglong granite in the study area (only 0.03–0.04%, according to Li et al. 2013). Mantle-derived fluids are added to the ore-forming fluids of the Jiaodong gold deposit (Fan et al. 2003; Deng et al. 2003; Deng et al. 2005; Mao et al. 2008). Therefore, the P in the gold-mineral microregions in this study may be derived from mantle-derived fluids.

The crust contains a higher abundance of P, which exists in the form of a major element (P_2O_5) (Zhang et al. 2012). However, P-rich fluid phases are unlikely to be differentiated during the evolution of magma-hydrothermal fluids in the shallow crust (Tang Yong et al. 2008). Additionally, although P is a compatible element, its compatibility is limited to the mantle; in granite, P is a trace element that is controlled by minerals, such as apatite—making it incompatible (Roilison 1993). P_2O_5 in Linglong granite in the study area is 0.03–0.04 (wt%), but the content in potassium feldspar, sericitised, and silicified and pyrite-sericite-quartz altered granite are ~0.01, ~0.025, and ~0.01, respectively. After alteration, the content is ~0.03, and P_2O_5 is significantly reduced; the minerals that control P (apatite, monazite, etc.) are gradually consumed (Li et al. 2013). Additionally, numerous peraluminous, quasi-aluminous, and other granite magma hydrothermal-evolution systems show that the

P_2O_5 content is controlled by the fractional crystallization of apatite (Li et al. 2007; Clemens et al. 2011). Therefore, the P in the ore-forming fluid is less likely to originate from the crust or wall rock differentiation fluid.

In summary, the gold-mineral microzone P may have originated from the deep mantle, indicating that mantle-derived fluids have been added to the ore-forming fluid.

(4) W

W has iron-philic properties (metal and silicate differentiation) but is a typical highly incompatible stone-philic element in the crust-mantle differentiation process and, therefore, enriched in the crust and obviously depleted in the mantle (Yin et al. 2002; Ma 2009).

Existing experimental tests show that under temperature and pressure conditions of 800 °C and 1.5 × 108 Pa, W is primarily concentrated in the melt, and the distribution coefficient in the fluid is very low (KD<0.3) (Xu et al. 1992). Therefore, the content of W in the deep mantle fluid is very small, and the W contained in gold minerals is unlikely to have originated in the mantle fluid.

To study the solubility of W in ore-forming fluids, numerous experiments have been conducted to detect the solubility of WO_3 in a high-temperature and high-pressure environment at temperatures of 600–250 °C and pressures of 100, 200, and 34 MPa. The results inferred that $Na^+(K^+)$ ions may be complexed with W in the high-temperature hydrothermal system, that alkaline conditions can promote the dissolution of WO_3 in the hydrothermal system, and that high-temperature conditions can significantly improve the ability of the ore-forming fluid to react with the metal ore-forming element tungsten (Wood and Vlassopoulos 1989; Gong et al. 2004). The existing W content data of Linglong granite in the northwestern part of Jiaodong is 0.33–1.65 μg/g, with an average of 0.77 μg/g (Gong et al. 2013). The ore-forming fluid in the study area was a medium-low temperature, low-salinity fluid with temperature and salinity that were significantly reduced after mineralisation. The solubility of W in the low-temperature and low-Na + fluid environments in the study area was low, and the W content in the surrounding rock was low. W is difficult to extract from the surrounding rock during water-rock reactions, and the average abundance of W in the crust is 1 μg/g (Rudnick and Gao 2003). In silicate magma, W can form WO coordination with O; thus, the solubility of W in silicate magma is very high (Ma 2009). The original magma has a higher temperature, the ore-forming fluid in the study area is a $CO_2 + H_2O + NaCl \pm CH_4$ system, and the solubility of WO_3 in the fluid is higher. Therefore, W can migrate to the original magmatic fluid and ultimately coexist with gold. From 130–110 Ma (Early Cretaceous), magma erupted, large-scale gold deposits formed in the Taihang, Luxi, Liaodong, and Dabie–Sulu areas, and the metallogenic age of the Jiaodong gold deposit was ~120 Ma (Li et al. 2013; Ma et al. 2014; Goldfarb and Santosh 2014). W may originate from silicate magmatic fluid that was differentiated during the Early Cretaceous magmatism.

Tungsten precipitation experiments and research on the formation mechanism of tungsten deposits reveal that the pH of ore-forming fluid is one of the principal factors controlling W precipitation and that the pH is relatively alkaline (>5.9) (Graupner et al. 1999). Elemental Ca, Mn, and Fe have a strong affinity for W and form an "enrichment barrier", which precipitates W, facilitating enrichment in an alkaline and high calcium ion concentration environment (Wang 1987). The principal gold-bearing mineral in the study area is iron-rich pyrite—Fe is also present in the study of gold-bearing minerals. Therefore, the inclusion of W into the gold particles reflects the slightly alkaline gold-forming environment.

No W was found in the gold particle microregions of the Linglong gold mine, which reflects that the precipitation fluid environment of the altered rock-type gold deposit is more alkaline than the quartz vein type.

(5) Other elements

Small amount of S, Fe, and Zn were in the single-particle microzone geochemistry of the silver-gold deposits in the study area. Because of the low content and the influence of co-existing gold pyrite (FeS_2) and sphalerite (ZnS), the elements may be mixed with gold-bearing minerals; thus, accurate determination of the source environment is impossible.

4.3.4 Sources of Ore-Forming Minerals and Fluids

The sources of fluids and ore-forming materials in the Xinli, Sanshandao, and Linglong gold deposits have been determined by studying numerous stable isotopes. In this study, the statistics of hydrogen and oxygen isotope data are available and based on magma water $\Delta D = -80‰$ to $-40‰$; $\delta^{18}O$ water $= 5.5–9.5‰$ (Sheppard 1986; Ohmoto 1986); and metamorphic water $\delta D = -65‰$ to $-20‰$ (Hugh and Taylor 1974). The δD values of the three typical deposits are relatively similar and concentrated in the range of the original magma water. The Linglong gold mine has a relatively wide $\delta^{18}O$ water, which tends to be close to atmospheric precipitation (Fig. 4.10). The Sanshandao gold mine is similar to the Xinli gold mine, which is primarily concentrated in the original magma water and a small area nearby and tends to approach atmospheric water downward. The overall display is dominated by primitive magma water with the addition of atmospheric precipitation in the later period. Compared with the Linglong gold mine, the amount of atmospheric precipitation in the study area is relatively small, which is consistent with the conclusion of N. Combined with the abovementioned analysis of micro-area elements, the ore-forming fluid of the Jiaodong gold deposit is a mixture of multiple sources and primarily composed of magmatic water, with the addition of mantle-derived fluids and mixed with atmospheric water (the altered rock type mixed in less). Numerous studies on the sources of ore-forming materials have revealed that the formation and intrusion of metamorphic rocks in the Archean and Proterozoic and granites in the Late Triassic and Late Jurassic occurred earlier than mineralisation, thus providing the possibility of ore formation (Mao et al. 2011; Li et al. 2013; Goldfarb and Santosh

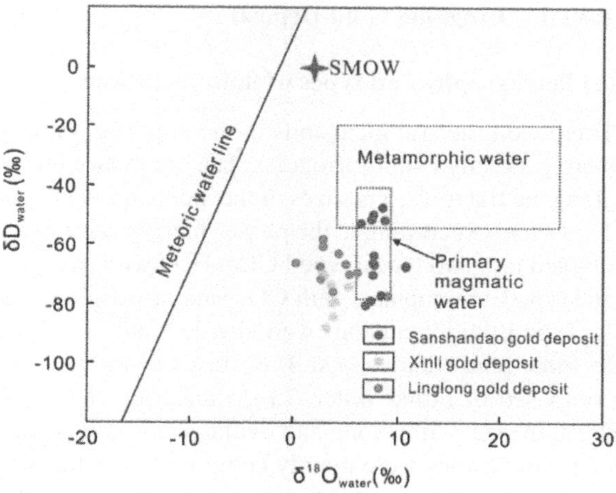

Fig. 4.10 Diagram showing $\delta^{18}O_{water}$(‰) versus δD (‰) for typical gold deposits in Jiaodong Penisula. *Data sources* Sanshandao gold deposit (Fan et al. 2003; Mao et al. 2008); Xinli gold deposit (Deng et al. 2015b); Linglong gold deposit (Mao et al. 2008; Wen et al. 2015)

2014; Wen et al. 2015). Early Cretaceous granites have similar intrusion periods to the gold deposits but are not directly related to the space within the formation of the ore body. Therefore, the Early Cretaceous granites are less likely to provide mineralisation (Goldfarb and Santosh 2014). Numerous studies have revealed that the formation of Early Cretaceous meso-basic dyke rocks in the Jiaodong area is closely related to gold deposits in time and space, with similar deep mantle source areas (Deng et al. 2003; Fan et al. 2003; Mao et al. 2008; Tan et al. 2012, 2015).

4.4 The Micro-Geochemical Characteristics of Pyrite in the Linglong-Type and Jiaojia-Type Gold Deposits in the Jiaodong Gold Province: Implications for genesis of gold deposit

4.4.1 Fluid Inclusions

To accurately demonstrate the changes in fluid properties, samples with fine veins were selected and milled to inclusion pieces that represented the ore-forming stage in both Jiaojia and Linglong gold deposits. The mineral compositions of the samples were simple and mainly composed of quartz, pyrite, and calcite. The fluid inclusions were mainly contained in the quartz. Samples from Linglong gold deposit included LLW50605-11, LLW50605-18, LLW50605-10, LLW50603-15, LLW50603-2, LLW50605-9, LLW50603-8, LLW50603-7, LLW50603-8, LLW50603-14, and LL206-5. The samples from the Jiaojia gold deposit were JJ60-3, JJ60-5, JJ230-1, and JJ60-1.

4.4.1.1 Linglong Gold Deposit

(1) Petrography and types of fluid inclusions

There were several inclusions in the ore. The primary inclusions were isolated and were generally oval or irregular. The secondary inclusions were linearly distributed along the fractures. The sizes of the inclusions were generally 4–11 μm, with a few at 11–18 μm. According to the phase changes at room temperature, the inclusions were divided into two types: type I CO_2–H_2O with two-phase and three-phase inclusions, and type II monophase with CO_2 vapour inclusions (Fig. 4.11a–d).

Type I fluid inclusions were divided into Ia CO_2–H_2O two-phase inclusions and Ib three-phase inclusions. Type Ia CO_2–H_2O two-phase inclusions were mainly composed of liquid water (L_{H2O}) and vapour CO_2 (V_{CO2}) and were generally 4–9 μm in size, with a long and ovular shape (Fig. 4.11a–b). Type Ib CO_2–H_2O three-phase inclusions were mainly composed of liquid water (L_{H2O}), liquid CO_2 (L_{CO2}), and vapour CO_2 (V_{CO2}) at room temperature. The inclusions had sizes of generally 5–18 μm, with a long and ovular shape (Fig. 4.11c).

Type II monophase vapour inclusions with CO_2 accounted for more than 90% of the total volume at room temperature. These conclusions have sizes of ≤6 μm and show mainly oval and circular shapes (Fig. 4.11d).

(2) Microthermometry

The temperature and salinity of the fluid inclusions in a total of nine samples of the Linglong gold deposit were analysed for each mineralisation stage. Based on the analysis data, the homogeneous temperature and salinity histogram of each stage are shown in Fig. 4.12. Combined with the microscopic analysis of the inclusions, the homogeneous temperature and salinity data of the fluid inclusions in each mineralisation stage are listed in Table 6.3.

In general, the homogeneous temperature range of the fluid inclusions from the Linglong gold deposit was 175.6–385.6 °C (average of ~273.8 °C), in which the temperature of the main mineralisation stage was 250–300 °C. The temperature and salinity at each stage were as follows.

The milky white quartz stage (1) mainly contained Type Ia CO_2–H_2O two-phase inclusions that were mainly composed of liquid water (L_{H2O}) and vapour CO_2 (V_{CO2}), with a small amount of Type Ib CO_2-containing three-phase inclusions. The homogeneous temperature range of the inclusions was 201.2–325.7 °C, concentrated between 250 and 280 °C. The salinity of the inclusion ranged from 4.1 to 12.3% NaCl.eqv., and was mainly concentrated between 8.0% and 10.0% NaCl.eqv.

The quartz-pyrite stage (2) was dominated by Type I CO_2–H_2O two-phase inclusions. The homogeneous temperature range of the inclusions was 198.6–378.4 °C, and concentrated between 240 and −270 °C. The salinity range of the inclusions was 4.4–10.1% NaCl.eqv., and was concentrated between 5.0 and −9.0% NaCl.eqv.

The polymetallic sulfide stage (3) was dominated by Type Ib CO_2–H_2O three-phase inclusions that were mainly composed of liquid water (L_{H2O}), liquid CO_2

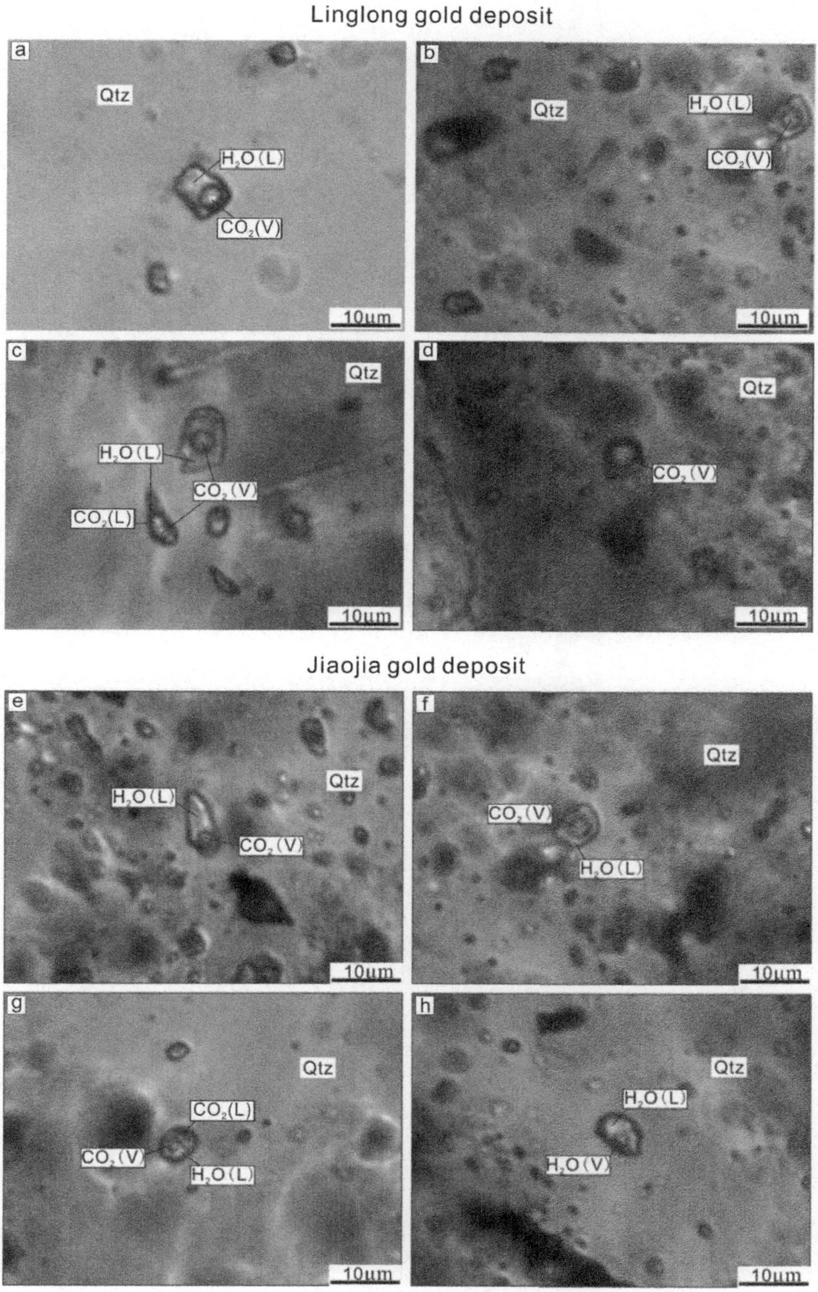

Fig. 4.11 Representative primary fluid inclusion assemblages from the Linglong gold deposit (**a**–**d**) and Jiaojia gold deposit (**e**–**h**). **a**–**b** Type I CO_2–H_2O two-phase inclusions, **c** type Ib CO_2–H_2O three-phase inclusions, and **d** type II monophase CO_2 vapour inclusions in the Linglong gold deposit; **e**–**f** type Ia CO_2–H_2O two-phase inclusions, **g** type Ib CO_2–H_2O three-phase inclusions, and **h** type III H_2O vapour-liquid two-phase inclusions in the Jiaojia gold deposit

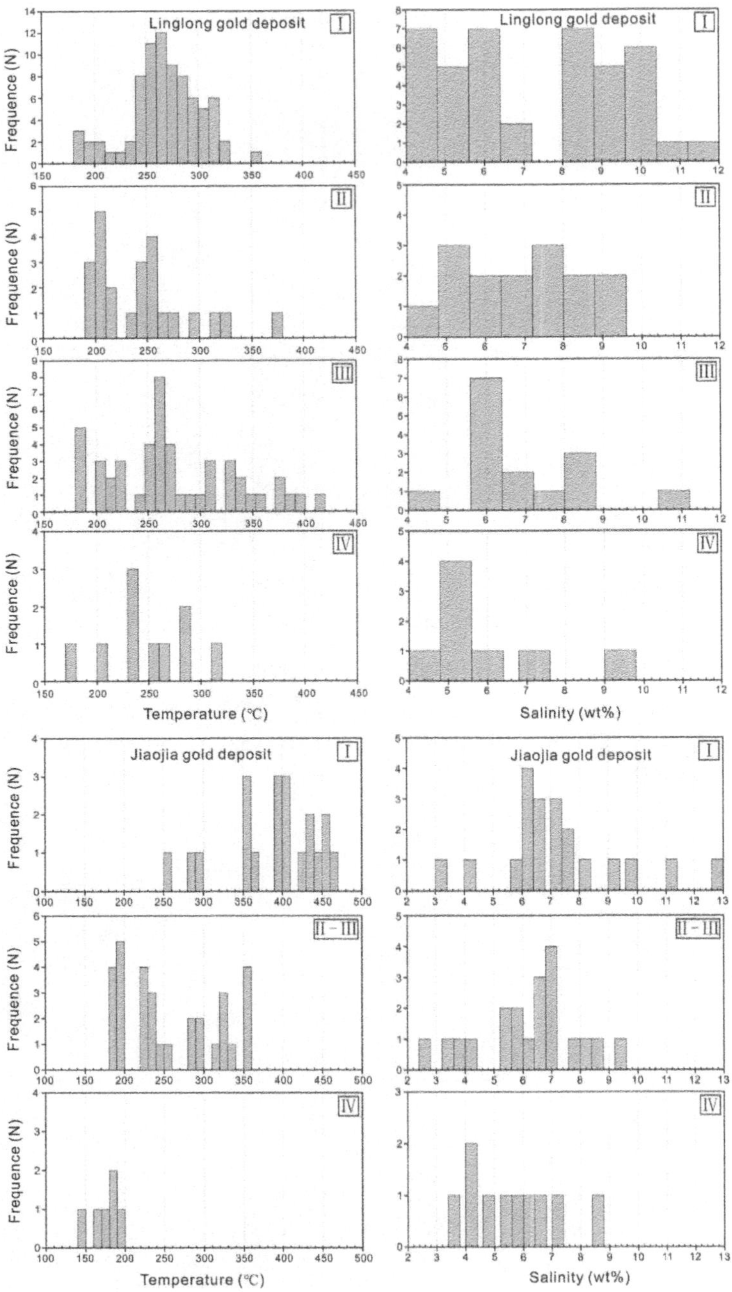

Fig. 4.12 Frequency histogram of total homogenisation temperatures and salinities of fluid inclusions from the four stages of the Linglong and Jiaojia gold deposits

(L_{CO2}), and vapour CO_2 (V_{CO2}). The homogeneous temperature range of the inclusions was 257.4–385.6 °C and mainly concentrated between 240 and −260 °C. The salinity range of the inclusions was 4.6–12.0.% NaCl.eqv., and concentrated between 5.5–6.5% NaCl.eqv.

The quartz–carbonate stage (4) was dominated by type II monophase CO_2 vapour inclusions. The homogeneous temperature range of the inclusions was 175.6–314.6 °C and concentrated between 230 and 250 °C. The salinity was mainly concentrated between 4.8 and 5.6% NaCl.eqv.

4.4.1.2 Jiaojia Gold Deposit

(1) Petrography and types of fluid inclusions

Inclusions in the Jiaojia gold deposit were relatively small, with most ranging from 4–16 μm. Based on the phase and composition of the inclusions, two types of inclusions were recognised in the Jiaojia gold deposit: type I CO_2–H_2O two-phase and three-phase inclusions, and type III H_2O vapour-liquid two-phase inclusions (Fig. 4.11e–h).

Type I fluid inclusions could also be divided into Ia CO_2–H_2O two-phase inclusions and Ib three-phase inclusions, similar to those of the Linglong gold deposit (Fig. 4.11e–g).

The type III H_2O vapour-liquid two-phase inclusions were composed mainly of two-phase water vapour (VH_2O) and liquid water (LH_2O). They were characterised by an oblong and elliptic shape with sizes of 5–8 μm (Fig. 4.11h).

(2) Microthermometry

Homogeneous temperature and salinity data of the five total samples from each mineralisation stage of the Jiaojia gold deposit are listed in Table 4.3 and Fig. 4.12.

The inclusions in stage (1) were Type I CO_2–H_2O two-phase. The homogeneous temperature range of the inclusions was 256.7–473.5 °C and was concentrated at 350–400 °C. The salinity of the inclusions ranged from 3.2 to 12.7% NaCl.eqv. and was concentrated between 6.0 and 7.0% NaCl.eqv.

The inclusions in stages (2) and quartz–polymetallic (3) were mainly the Ia CO_2–H_2O two-phase and Ib three-phase types. The homogeneous temperature range of the inclusions was 164.2–356.2 °C, with primary concentrations at 280–340 °C and 170–240 °C. The salinity of the inclusions ranged from 2.6 to 9.5% NaCl.eqv. and was concentrated between 6.0 and 7.0% NaCl.eqv.

The inclusions in the carbonate stage (4) were mainly type III H_2O vapour-liquid two-phase inclusions. The homogeneous temperature range of the inclusions was 139.1–194.3 °C and was concentrated between 160–200 °C. The salinity range of the inclusions was 3.5%–8.6% NaCl.eqv., with a concentration between 3.5–5.0% NaCl.eqv.

Table 4.3 Microthermometric data of fluid inclusions at the Linglong and Jiaojia gold deposits

Name	Stage	Sample	$T_{m\text{-}ice}$(°C)	Average (°C)	$T_{h\text{-}tot}$ (°C)	Average (°C)	Salinity (wt%NaCleqv)	Average (wt%NaCleqv)
Linglong	Stage I	LL206-1	−3.8 ~ −0.3	−1.9	201.2–254.2	255.4	4.1–12.3	8.1
		LL206-5	−5.7 ~ 2.3	−2.4	241.3–288.7	257.4	4.3–8.7	6.3
		LLW-7	−6.1 ~ 9.3	2.7	254.3–325.7	294.2	8.3–9.2	8.7
		LLW-8	−8.4 ~ 8.1	−1.7	237.2–324.7	282.2	4.3–11.8	7.2
	Stage II	LL206-3	−4.5 ~ 2.4	−1.9	198.6–291.4	232.4	5.2–10.1	7.2
		LLW-15	−3.4 ~ 9.1	4	198.6–378.4	267.2	4.4–5.5	4.9
	Stage III	LLW-14	−3.7 ~ 7.5	−1.4	257.4–329.7	286.1	4.6–8.1	6.2
		LLW-11	−8.5 ~ −1.7	−5.2	265.4–385.6	339.8	7.2–12	9.6
	Stage IV	LLW-18	−6.2 ~ 4.2	0.01	175.6–314.6	249.4	4.1–9.3	5.7
Jiaojia	Stage I	JJ60-3	−2.1 ~ −3.4	−2.8	256.7–408.3	389.5	3.2–11.9	7.8
		JJ60-5	−5.8 ~ 7.9	−0.7	334.3–473.5	396.3	4.4–12.7	8.5
	Stage II-III	JJ60-1	−3.4 ~ −1.2	−2.1	164.2–335.2	255.8	2.6–9.3	6.2
		JJ230-1	−3.6 ~ −1.2	−2.5	184.3–356.2	243.2	3.6–9.5	6.4
	Stage IV	JJ54-1	−3.5 ~ −2.4	−2.8	139.1–194.3	170.1	3.5–8.6	5.6

$T_{m\text{-}ice}$, final melting temperature of water ice; $T_{h\text{-}tot}$, temperature of total homogenization of the inclusions; wt% NaCl equiv., weight percent NaCl equivalent

4.4.2 Geochemical Composition of Pyrite

(1) Major and trace element compositions

In this study, previous electronic probe data for major elements of gold-bearing pyrite from the Jiaojia-type and Linglong-type gold deposits in the Jiaodong Peninsula are listed in Table 4.4. The major elemental compositions of pyrite were Fe and S.

The trace elements of pyrite mainly included Cu, Pb, Zn, Au, Ag, As, Co, Ni, Sb, Bi, Se, Te, Mo, Sn, Ti, Cr, and Hg (Table 4.5). The continental crust-normalised trace element diagram shows that pyrites from both the Jiaojia and Linglong gold deposits had similar features, were enriched in Cu, Pb, and Zn, and were depleted in Nb, Ta, Zr, and Hf (Fig. 4.13a).

The chondrite-standardised rare earth element diagram of pyrite shows enrichment of light rare earth elements (LREE) (Fig. 4.13b). Pyrites of the Jiaojia gold deposit had ΣREE values of 0.66×10^{-6} to 50.8×10^{-6} (average value of 27.8×10^{-6}), LREE/HREE ratios of 2.99-18.3 (average value of 10.3), $(La/Yb)_N$ values of 1.73–35.9 (average value of 14.3), δEu values of 0.58–1.02 (average value of 0.87), and δCe values of 1.03–1.22 (average value of 1.09). The Linglong gold deposit pyrites had ΣREE values of 2.59×10^{-6} to 45.8×10^{-6} (average value of 20.1×10^{-6}), LREE/HREE ratios of 2.17–33.7 (average value of 18.2); $(La/Yb)_N$ ratios of 2.08–96.5 (average value of 43.6), δEu values of 0.97–1.97 (average value of 1.34), and δCe values of 1.0–1.03 (average value of 1.01).

(2) Sr-Nd isotope compositions

The Sr-Nd isotope compositions of 14 pyrites in the Linglong and Jiaojia gold deposits were analysed in this study (Table 4.6). Pyrites in the Linglong gold deposit had $(^{87}Sr/^{86}Sr)_i$ values of 0.71215–0.71331 (average value of 0.713) and $\varepsilon_{Nd}(t)$ values of −20.56 to −12.73 (average value of −17.43). Pyrites in the Jiaojia gold deposit had $(^{87}Sr/^{86}Sr)_i$ values of 0.71075–0.71125 (average value of 0.711) and ε_{Nd} (t) values of −20.46 to −9.83 (average value of −13.92) (Fig. 4.14). Previous studies on the Sr-Nd isotope compositions of pyrites in the Linglong gold deposit and Early Cretaceous mafic dykes, Linglong granites, and Guojialing granodiorites are listed in Table 6.6.

4.4.2.1 S Isotope Composition

The S isotope compositions of 14 pyrites in Linglong and Jiaojia gold deposits were analysed in this study (Table 4.7). The value of $\delta^{34}S$ in the Linglong gold deposit was 6.27–9.80‰, with an average value of 8.31‰. The $\delta^{34}S$ values of the pyrite from stage 1 were 7.16–9.39‰ (average value is 8.58‰); stage 2 was 6.27–9.80‰ (average value of 6.87‰), and stage 3 was 6.98–9.60‰ (average value of 8.34‰). The $\delta^{34}S$ value of the Jiaojia gold deposit was 8.43‰–11.2‰, with an average value of 10.23‰; the $\delta^{34}S$ values of pyrite from stage 1 were 9.46–11.2‰ (the average

Table 4.4 Major elements of pyrite in Linglong and Jiaojia gold deposits

Type	Name	Fe (%)	S (%)	δFe	δS
Linglong-type gold deposit	Linglong Dongshan	46.92	50.79	0.79	−4.98
		47.56	51.49	2.17	−3.67
		47.32	49.36	1.65	−7.65
		47.32	49.22	1.65	−7.91
		47.36	51.11	1.74	−4.38
	Linglong Xishan	46.31	53.61	−0.52	0.3
		45.83	52.18	−1.55	−2.38
		46.72	51.32	0.37	−3.99
	Denggezhuang	46.22	52.26	−0.71	−2.23
		46.39	52.84	−0.34	−1.14
		46.31	52.93	−0.52	−0.97
		46.75	52.63	0.43	−1.53
		46.65	53.13	0.21	−0.6
		46.77	52.75	0.47	−1.31
	Rushan	45.31	52.34	−2.66	−2.08
		45.93	53.73	−1.33	0.52
		45.09	53.35	−3.14	−0.19
		46.92	52.11	0.79	−2.51
		45.92	53.24	−1.35	−0.39
		46.07	53.32	−1.03	−0.24
Jjiaojia-type gold deposit	Lingshangou	46.66	52.23	0.24	−2.28
		45.55	53.05	−2.15	−0.75
		46.11	52.15	−0.95	−2.43
		45.54	51.61	−2.17	−3.44
	Xincheng	43.44	51.92	−6.68	−2.86
		45.79	52.74	−1.63	−1.33
		44.53	52.55	−4.34	−1.68
		44.31	52.21	−4.81	−2.32
		44.63	52.04	−4.12	−2.64
		44.39	51.38	−4.64	−3.87
	Jiehe	44.67	51.26	−4.04	−4.1
		45.61	52.35	−2.02	−2.06
		45.36	52.07	−2.56	−2.58
		44.28	50.93	−4.88	−4.71
		43.61	49.79	−6.32	−6.85
	Jiaojia	43.92	50.3	−5.65	−5.89

(continued)

Table 4.4 (continued)

Type	Name	Fe (%)	S (%)	δFe	δS
		44.64	51.27	−4.1	−4.08
		44.85	51.04	−3.65	−4.51
		46.29	53.02	−0.56	−0.8
		46.29	51.41	−0.56	−3.82

value of 10.7‰) while those from stages 2 and 3 were 8.43–10.8‰ (average value of 10.6‰) and 8.89–10.83‰ (average value of 10.5‰), respectively.

In addition, we collected S isotope data from previous studies on the pyrite in Linglong-type Rushan and Denggezhuang gold deposits the Jiaojia-type Sanshandao gold deposit, the Jiaodong and Jingshan groups, medium-basic dykes, and Linglong, Yanjiahe, and Guojialing granite (Fig. 4.15 and Table 4.8).

4.4.3 Features and Sources of Ore-Forming Fluids

In this study, the ore-forming fluid inclusions of the Linglong and Jiaojia gold deposits displayed similar types and petrography (Fig. 4.11). Quartz in the early and main mineralisation stage mainly contained type I CO_2-rich two-phase (type Ia L_{H2O} + V_{CO2}) and/or three-phase inclusions (type Ib L_{H2O} + L_{CO2} + V_{CO2}). However, in the late stage, the Linglong gold deposit contained type II monophase CO_2 vapour inclusions, while the Jiaojia gold deposit had type III H_2O vapour-liquid two-phase inclusions. The ore-forming fluids had a medium-high temperature, were CO_2-rich, and displayed a medium-low salinity. In addition, the temperatures of the ore-forming fluids decreased from the early to late stages in the Jiaojia gold deposit (Fig. 4.12). However, the temperatures of the ore-forming fluids weakly increased in the main mineralisation stage and then decreased in the late stage in the Linglong gold deposit (Fig. 4.12).

Because the trace element compositions of pyrites are constrained by ore-forming fluids and rare earth elements (REEs) are relatively inactive elements in the hydrothermal systems, the REEs of pyrites were used to track the source of ore-forming fluids and water-rock interactions in previous studies (Moh 1980; Henderson 1984). REEs can be combined with halogen elements as well as CO_3^{2-} and SO_4^{2-} in a solution, thus forming chlorides, fluorides, sulfates, carbonates, or other complexes (Chen et al. 2014; Li et al. 2020). In this study, pyrite of the Jiaojia gold deposit showed low ΣREE contents of 0.66×10^{-6}-50.81×10^{-6} and LREE enrichment with LREE/HREE ratios of 2.99–18.32 and $(La/Yb)_N$ values of 1.73–35.9 (Fig. 4.13b). Pyrite of the Linglong gold deposits also showed low ΣREE contents of 2.59×10^{-6} to 45.8×10^{-6}, LREE enrichment with LREE/HREE ratios of 2.17–33.68, and $(La/Yb)_N$ ratios of 2.08–96.47 (Fig. 4.13b). These features indicated that the ore-forming fluid contained a large amount of Cl^- or F^- (Haas et al. 1995; Keppler

Table 4.5 Trace elements compositions of pyrites in the Linglong and Jiaojia gold deposits

Location	Jiaojia				Linglong		
Sample no.	JJ60-1	JJ60-2	JJ230-2	JJ230-6	LL206-2	LL206-3	LL206-6
Li	2.11	2.54	2.97	2.65	2.71	2.77	2.75
Be	0.09	0.12	0.10	0.02	0.04	0.04	0.26
Sc	0.20	0.26	0.32	0.03	0.04	0.12	0.10
V	1.53	1.36	2.43	0.33	0.50	1.77	1.82
Cr	1.36	0.90	9.18	0.30	0.23	0.85	1.07
Hg	0.23	0.25	0.31	0.35	1.72	2.18	3.24
Co	22.91	33.35	281.91	45.19	10.02	1.03	0.51
Ni	42.82	47.61	553.25	31.79	0.75	6.29	6.44
Cu	72.70	76.47	98.10	5.62	18.43	32.60	7.09
Zn	79.91	239.56	81.47	8.40	215.12	39.57	15.63
Ga	0.92	1.00	0.76	0.42	0.42	0.95	0.67
Rb	2.67	2.89	1.76	0.45	0.18	2.58	2.32
Sr	7.15	25.34	9.09	2.45	15.00	110.50	18.34
Zr	70.22	85.37	59.76	0.69	0.31	5.28	12.57
Nb	1.11	1.99	1.09	0.11	0.08	0.76	1.42
Mo	6.78	4.70	2.87	0.28	0.71	0.40	0.48
Cd	0.47	1.25	0.44	0.09	0.86	0.22	0.11
In	0.01	0.01	0.01	0.00	0.02	0.07	0.01
Cs	0.09	0.19	0.06	0.01	0.04	0.03	0.06
Ba	36.93	601.27	167.20	23.67	114.55	660.87	131.27
Hf	2.39	2.25	1.76	0.04	0.01	0.15	0.44
Ta	0.07	0.16	0.06	0.01	0.01	0.05	0.17
Pb	783.81	917.15	451.17	465.37	138.14	91.16	44.47
Bi	57.28	54.61	54.40	135.70	4.03	4.23	1.97
Th	2.93	3.07	2.58	0.18	0.26	1.50	1.85
U	172.74	374.82	435.92	84.45	311.06	84.57	31.73
La	6.26	7.69	10.96	0.1	0.41	3.02	11.66
Ce	11.81	14.25	22.24	0.24	0.78	5.47	21.39
Pr	1.22	1.44	2.54	0.02	0.09	0.58	2.21
Nd	4.7	4.72	10.33	0.1	0.34	1.85	7.76
Sm	0.83	0.78	1.72	0.02	0.11	0.25	1.15
Eu	0.14	0.27	0.4	0.01	0.04	0.17	0.32
Gd	0.64	0.81	1.19	0.03	0.16	0.29	0.72
Tb	0.09	0.1	0.14	0.01	0.03	0.02	0.06
Dy	0.64	0.69	0.63	0.04	0.25	0.12	0.28

(continued)

Table 4.5 (continued)

Location	Jiaojia				Linglong		
Sample no.	JJ60-1	JJ60-2	JJ230-2	JJ230-6	LL206-2	LL206-3	LL206-6
Ho	0.11	0.13	0.1	0.01	0.04	0.02	0.04
Er	0.42	0.48	0.28	0.03	0.16	0.07	0.11
Tm	0.07	0.08	0.03	0	0.02	0.01	0.01
Yb	0.43	0.61	0.22	0.04	0.14	0.07	0.09
Lu	0.1	0.09	0.03	0	0.02	0.01	0.02
Y	3.83	5.56	3.43	0.25	1.72	0.67	0.68
ΣREE	27.45	32.13	50.81	0.66	2.59	11.94	45.8
LREE	24.96	29.14	48.18	0.5	1.77	11.34	44.48
HREE	2.49	2.99	2.63	0.17	0.82	0.6	1.32
LREE/HREE	10.02	9.75	18.32	2.99	2.17	18.87	33.68
LaN/YbN	10.37	9.1	35.9	1.73	2.08	32.18	96.47
δEu	0.58	1.02	0.86	1	0.97	1.97	1.07
δCe	1.05	1.05	1.03	1.22	1	1.01	1.03

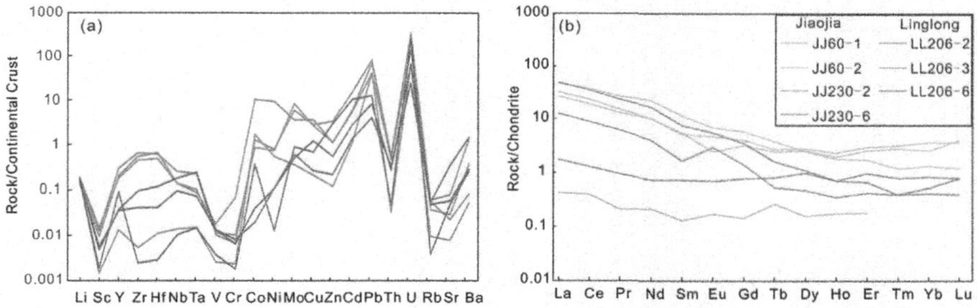

Fig. 4.13 **a** Continental crust-normalised immobile trace element patterns and **b** chondrite-normalised REE patterns of pyrite from the Linglong and Jiaojia gold deposits. Chondrite and primitive mantle values are from Sun and McDonough (1989); continental crust values are from Rudnick and Gao (2003)

1996; Bi et al. 2004; Li et al. 2020). Moreover, REEs and HFESEs always exist in fluid inclusions or crystal defects of pyrites; thus, the compositions of trace elements of pyrite represent the composition of ore-forming fluids (Bi et al. 2004). Oreskes and Einaudi (1990) have reported that rich-Cl hydrotherms have low Hf/Sm, Th/La, and Nb/La ratios (<1), and rich-F hydrotherms are the opposite (>1). Linglong gold deposits had a Th/La ratio of 0.159–0.634 and a Nb/La ratio of 0.122–0.252, which implied a rich Cl^- fluid. The Jiaojia deposits contained a large Th/La range of 0.235–1.80 and a similarly large Nb/La range of 0.099–1.10, thereby indicating fluids rich in Cl^- and F^-.

Table 4.6 Sr-Nd isotopes of pyrite in Jiaodong gold deposits and Late Mezosoic magmatic rocks

Location	Sample No.	Type	Rb (ppm)	Sr (ppm)	$^{87}Rb/^{86}Sr$	$^{87}Sr/^{86}Sr$	±2σ	$(^{87}Sr/^{86}Sr)_i$	Sm (ppm)	Nd (ppm)	$^{147}Sm/^{144}Nd$	$^{143}Nd/^{144}Nd$	±2σ	$\varepsilon_{Nd}(t)$	Data sources
Jiaojia	JJ60-2	Pyrite	2.88	25.34	0.3301	0.711817		0.71125	0.7807884	4.7158813	0.1001	0.511514		−20.46	This study (unpublished)
	JJ230-2		1.75	9.09	0.5601	0.712062		0.71111	1.7158051	10.327624	0.1004	0.512059		−9.83	
	JJ230-6		0.44	2.45	0.5285	0.711653		0.71075	0.0190275	0.1041507	0.1105	0.511983		−11.47	
Linglong	LL206-2		0.17	15.00	0.0344	0.712205		0.71215	0.1138549	0.3385421	0.2033	0.511991		−12.73	
	LL206-3		2.58	110.50	0.0677	0.712277		0.71216	0.2475256	1.845392	0.0811	0.511574		−19	
	LL206-6		2.31	18.33	0.3662	0.713937		0.71331	1.1508364	7.7584553	0.0897	0.5115		−20.56	
Linglong gold deposit	98DK03	Pyrite	1.10	1.61	1.98	0.714932	0.000016	0.711470	0.062000	0.352000	0.106200	0.511577	0.000018	−19.30	Yang and Zhou (2001)
	98DK07		0.37	0.20	5.25	0.719975	0.000035	0.710790	0.099000	0.627000	0.095800	0.511842	0.000018	−13.90	
	98JQ04		0.31	0.05	18.84	0.743299	0.000048	0.710370	0.019000	0.088000	0.133300	0.511989	0.000075	−11.60	
	98JQ07		0.52	0.07	21.15	0.745900	0.000020	0.708930	0.058000	0.398000	0.087900	0.511663	0.000015	−17.30	
	98JQ08		0.58	0.09	19.61	0.743007	0.000032	0.708720	0.080000	0.527000	0.091500	0.511510	0.000011	−20.30	
	98LL06		0.02	0.04	1.63	0.713005	0.000050	0.710160	0.044000	0.334000	0.079000	0.511622	0.000017	−18.00	
JIaojiao gold deposit	JJLT-01	Mafic dykes	102.00	1004.00	0.29	0.709539	0.000008	0.709034	7.100000	45.000000	0.095000	0.511792	0.000002	−14.93	Ma et al. (2014)
	JJLT-02		54.00	1481.00	0.11	0.709708	0.000005	0.709527	7.800000	45.000000	0.104000	0.511835	0.000006	−14.23	
	JJLT-03		105.00	952.00	0.32	0.709886	0.000006	0.709338	8.900000	51.000000	0.105000	0.511770	0.000001	−15.52	
	JJLT-04		126.00	958.00	0.38	0.710339	0.000004	0.709685	8.700000	51.000000	0.103000	0.511818	0.000005	−14.56	
	JJLT-05		73.00	1217.00	0.18	0.709517	0.000006	0.709217	6.900000	40.000000	0.104000	0.511833	0.000001	−14.27	
	JJLT-06		75.00	3162.00	0.07	0.709722	0.000003	0.709604	7.400000	45.000000	0.099000	0.511847	0.000001	−13.92	
	JJLT-07		78.00	934.00	0.24	0.709739	0.000004	0.709322	8.100000	50.000000	0.097000	0.511802	0.000001	−14.77	
Beibo gold deposit	B6-7		6.02	390.00	0.04	0.709672	0.000011	0.709593	3.090000	11.600000	0.161000	0.511841	0.000012	−15.00	Liu et al. (2009)
	B6-6		12.80	364.00	0.10	0.709601	0.000014	0.709421	3.150000	11.900000	0.160700	0.511849	0.000010	−14.80	
	B6-5		82.80	1026.00	0.23	0.709895	0.000010	0.709474	8.360000	51.500000	0.098100	0.511788	0.000090	−15.00	
	B6-4		83.90	1005.00	0.24	0.709924	0.000012	0.709489	8.330000	51.800000	0.097200	0.511784	0.000090	−15.10	
	B6-3		85.70	1139.00	0.22	0.709802	0.000013	0.709422	7.970000	50.100000	0.096200	0.511785	0.000010	−15.10	
	B6-2		78.50	1161.00	0.20	0.709791	0.000010	0.709450	8.130000	50.200000	0.097900	0.511781	0.000011	−15.20	

(continued)

Table 4.6 (continued)

Location	Sample No.	Type	Rb (ppm)	Sr (ppm)	$^{87}Rb/^{86}Sr$	$^{87}Sr/^{86}Sr$	±2σ	$(^{87}Sr/^{86}Sr)_i$	Sm (ppm)	Nd (ppm)	$^{147}Sm/^{144}Nd$	$^{143}Nd/^{144}Nd$	±2σ	$\varepsilon_{Nd}(t)$	Data sources
	B6-1		76.30	1093.00	0.20	0.709803	0.000010	0.709447	8.180000	49.600000	0.099700	0.511774	0.000080	−15.30	
	B5-2		78.30	1098.00	0.21	0.709811	0.000011	0.709448	8.410000	51.200000	0.099300	0.511779	0.000080	−15.20	
	B5-1		73.20	1559.00	0.14	0.710323	0.000008	0.710082	7.030000	43.400000	0.097900	0.511846	0.000090	−13.90	
	B4-3		101.00	798.00	0.37	0.710318	0.000010	0.709668	6.960000	42.100000	0.099900	0.511843	0.000010	−14.00	
Zhaoyuan area	LK-3-1		44.40	1025.90	0.12	0.709138	0.000007	0.708925	10.621248	71.472148	0.089907	0.511765	0.000005	−15.39	Deng et al. (2017)
	LK-3-3		43.30	1026.60	0.12	0.709078	0.000006	0.708870	10.905040	72.029600	0.091595	0.511757	0.000005	−15.57	
	LK-3-4		43.90	1031.10	0.12	0.709142	0.000006	0.708932	10.534914	70.091892	0.090932	0.511768	0.000006	−15.36	
	LK-6-1		52.70	983.50	0.15	0.710121	0.000007	0.709858	7.627025	48.113424	0.095906	0.511705	0.000005	−16.65	
	LK-6-3		47.70	953.80	0.14	0.710065	0.000007	0.709819	7.489491	49.372400	0.091775	0.511714	0.000004	−16.41	
	LK-6-4		49.20	940.20	0.15	0.710114	0.000007	0.709857	7.648194	49.069782	0.094297	0.511709	0.000005	−16.56	
	LK-6-5		50.40	955.30	0.15	0.710093	0.000008	0.709833	7.784784	50.006880	0.094183	0.511710	0.000004	−16.54	
	LK-7-1		62.70	961.80	0.19	0.709728	0.000006	0.709407	5.143836	33.855356	0.091921	0.511621	0.000005	−18.24	
	LK-7-2		60.30	943.40	0.18	0.709588	0.000008	0.709274	5.581494	35.495680	0.095133	0.511610	0.000004	−18.51	
	LK-7-3		61.40	944.60	0.19	0.709573	0.000007	0.709253	5.514016	36.551664	0.091268	0.511620	0.000006	−18.25	
	LK-9-1		52.90	742.60	0.21	0.709738	0.000007	0.709387	5.125733	32.573338	0.095203	0.511658	0.000006	−17.57	
	LK-9-2		52.40	753.90	0.20	0.709789	0.000008	0.709447	5.187325	33.256224	0.094368	0.511648	0.000006	−17.75	
	LK-9-3		57.30	799.50	0.21	0.709781	0.000009	0.709428	5.438783	34.292808	0.095952	0.511654	0.000006	−17.65	
	LK-10-1		65.10	1372.10	0.14	0.709753	0.000008	0.709519	13.938000	93.899700	0.089803	0.511678	0.000006	−17.09	
	LK-10-2		79.40	1227.50	0.19	0.709838	0.000009	0.709519	13.923312	92.744616	0.090826	0.511673	0.000006	−17.21	
	ZY-2-1		31.10	695.70	0.13	0.709534	0.000009	0.709313	5.038824	31.260120	0.097520	0.511676	0.000004	−17.25	
	ZY-2-2		31.50	712.50	0.13	0.709524	0.000007	0.709307	5.006451	31.620224	0.095790	0.511677	0.000005	−17.20	
	ZY-2-3		28.90	681.20	0.12	0.709546	0.000007	0.709337	5.112627	32.031038	0.096567	0.511694	0.000006	−16.89	
	ZY-5-1		64.70	813.20	0.23	0.710339	0.000009	0.709947	7.784381	50.184240	0.093845	0.511701	0.000004	−16.71	
	ZY-5-2		87.10	855.80	0.29	0.710353	0.000007	0.709852	7.424730	48.268848	0.093061	0.511701	0.000004	−16.70	
	ZY-5-3		79.00	812.60	0.28	0.710401	0.000009	0.709922	7.544763	48.481056	0.094152	0.511692	0.000005	−16.89	
	ZY-7-1		72.90	1024.40	0.21	0.710181	0.000007	0.709830	8.109543	51.024402	0.096155	0.511694	0.000005	−16.88	

(continued)

Table 4.6 (continued)

Location	Sample No.	Type	Rb (ppm)	Sr (ppm)	$^{87}Rb/^{86}Sr$	$^{87}Sr/^{86}Sr$	±2σ	$(^{87}Sr/^{86}Sr)_i$	Sm (ppm)	Nd (ppm)	$^{147}Sm/^{144}Nd$	$^{143}Nd/^{144}Nd$	±2σ	$\varepsilon_{Nd}(t)$	Data sources
	ZY-7-2		66.70	1038.40	0.19	0.710147	0.000008	0.709830	7.827110	49.817184	0.095056	0.511699	0.000005	−16.76	
	ZY-7-3		74.20	1044.90	0.20	0.710181	0.000007	0.709832	8.358275	52.972266	0.095460	0.511691	0.000004	−16.93	
	ZY-1-1		77.80	712.20	0.32	0.709567	0.000008	0.709029	7.822078	49.227948	0.096132	0.511728	0.000005	−16.21	
	ZY-1-2		73.80	772.20	0.28	0.709474	0.000009	0.709004	6.511225	40.650750	0.096906	0.511711	0.000005	−16.57	
Laiyang raea	LY-1-1		44.10	727.20	0.18	0.709156	0.000007	0.708858	7.842640	51.656800	0.091852	0.511625	0.000005	−18.16	Liang et al. (2019)
	LY-1-2		55.20	1177.40	0.14	0.709518	0.000008	0.709287	8.120841	52.311658	0.093920	0.511633	0.000005	−18.03	
	LY-1-5		57.90	883.60	0.19	0.709002	0.000009	0.708680	8.280544	54.985140	0.091111	0.511697	0.000005	−16.74	
	LY-2-3		52.20	763.20	0.20	0.708782	0.000011	0.708446	8.436725	53.476304	0.095448	0.511913	0.000006	−12.59	
	LY-2-4		58.30	820.00	0.21	0.708742	0.000007	0.708392	8.641960	55.672964	0.093912	0.511774	0.000004	−15.29	
Jiaobei terrana	08G01	Linglong granite	124.70	344.90	1.05	0.712054	0.000014	0.709672	1.400000	7.300000	0.113700	0.511652	0.000011	−17.60	Yang et al. (2012)
	08G02		172.50	359.50	1.39	0.712197	0.000013	0.709037	1.200000	6.300000	0.110400	0.511643	0.000013	−17.70	
	08G03		123.30	479.40	0.75	0.713781	0.000012	0.712086	3.300000	22.100000	0.089700	0.511456	0.000012	−20.90	
	08G04		72.00	993.40	0.21	0.712334	0.000014	0.711856	1.600000	11.700000	0.083500	0.511446	0.000021	−20.90	
	08G05		141.70	372.30	1.10	0.714291	0.000014	0.711783	1.400000	6.400000	0.134800	0.511471	0.000012	−21.50	
	08G19		98.50	957.10	0.30	0.709010	0.000012	0.708333	1.800000	13.300000	0.083300	0.511446	0.000014	−20.90	
	08G20		153.60	417.70	1.07	0.714551	0.000013	0.712128	2.500000	9.400000	0.158600	0.511519	0.000016	−21.10	
	08G40		108.30	668.80	0.47	0.713194	0.000014	0.712128	3.200000	24.300000	0.079800	0.511437	0.000011	−21.10	
	08G42		90.10	702.80	0.37	0.713263	0.000013	0.712418	2.600000	18.900000	0.083900	0.511421	0.000011	−21.40	
	08G50		128.20	291.20	1.28	0.714697	0.000013	0.711795	2.100000	8.400000	0.148400	0.511509	0.000011	−21.10	
	08G52		70.00	976.80	0.21	0.711753	0.000014	0.711281	2.000000	13.200000	0.091300	0.511447	0.000013	−21.10	
	08G53		74.70	916.60	0.24	0.712239	0.000015	0.711703	3.700000	25.000000	0.088600	0.511420	0.000013	−21.60	
	08G15		149.70	376.90	1.15	0.714674	0.000014	0.712058	0.800000	3.300000	0.148100	0.511593	0.000016	−19.40	

(continued)

Table 4.6 (continued)

Location	Sample No.	Type	Rb (ppm)	Sr (ppm)	$^{87}Rb/^{86}Sr$	$^{87}Sr/^{86}Sr$	±2σ	$(^{87}Sr/^{86}Sr)_i$	Sm (ppm)	Nd (ppm)	$^{147}Sm/^{144}Nd$	$^{143}Nd/^{144}Nd$	±2σ	$\varepsilon_{Nd}(t)$	Data sources
Jiaojia gold deposit	JJLL-1		65.71	430.00	0.44	0.712556	0.000005	0.711564	1.850000	8.970000	0.124300	0.511599	0.000007	−18.82	Ma et al. (2013)
	JJLL-2		54.11	438.00	0.36	0.710536	0.000006	0.709733	2.080000	8.660000	0.144900	0.511619	0.000003	−18.84	
	JJLL-3		66.33	602.00	0.32	0.712733	0.000005	0.712017	1.340000	6.860000	0.117700	0.511516	0.000012	−20.30	
	JJLL-4		48.44	614.00	0.23	0.710246	0.000008	0.709733	0.920000	6.330000	0.087800	0.511620	0.000007	−17.67	
	JJLL-5		86.42	423.00	0.59	0.712410	0.000010	0.711082	1.690000	9.710000	0.105200	0.511526	0.000009	−19.85	
	JJLL-6		92.40	352.00	0.76	0.713316	0.000013	0.711609	2.380000	14.330000	0.106700	0.511616	0.000005	−18.13	
	JJLL-7		84.36	268.00	1.45	0.715782	0.000005	0.712514	0.770000	5.450000	0.085400	0.511551	0.000010	−18.97	
	JJLL-8		82.41	348.00	0.69	0.712171	0.000006	0.710631	1.240000	9.550000	0.078500	0.511489	0.000002	−20.04	
	JJLL-9		86.84	215.00	1.17	0.712577	0.000012	0.709953	1.930000	13.020000	0.089600	0.511571	0.000016	−18.66	
Jiaobei terrana	06SD36	Guojialing granodiorite	84.50	2009.00	0.12	0.710840	0.000013	0.710600	10.700000	71.100000	0.090800	0.511801	0.000018	−14.70	Zhang et al. (2010)
	06SD47		89.80	962.00	0.27	0.711254	0.000012	0.710800	2.930000	17.700000	0.100300	0.511953	0.000008	−11.90	
	06SD64		70.50	1021.00	0.20	0.712469	0.000012	0.712100	1.350000	8.810000	0.092500	0.511413	0.000007	−22.30	
	06SD73		187.00	42.90	12.69	0.840923	0.000018		4.950000	22.800000	0.130900	0.511784	0.000008	−15.40	
	08G08		84.10	1109.20	0.22	0.711106	0.000013	0.710700	4.000000	25.700000	0.094800	0.511948	0.000013	−11.80	Yang et al. (2012)
	08G25		71.20	1744.90	0.12	0.711417	0.000012	0.711199	8.500000	58.200000	0.088400	0.511715	0.000014	−16.20	
	08G29		66.70	1708.30	0.11	0.711237	0.000012	0.711028	8.400000	89.600000	0.056900	0.511777	0.000013	−14.50	
	08G30		49.80	1214.30	0.12	0.711620	0.000013	0.711401	1.900000	10.500000	0.107800	0.511801	0.000012	−14.90	
	08G32		84.20	895.30	0.27	0.712066	0.000013	0.711563	3.400000	21.000000	0.098900	0.511692	0.000014	−16.80	
	08G37		76.10	1025.50	0.21	0.711985	0.000015	0.711588	1.900000	11.700000	0.096900	0.511727	0.000019	−16.10	
	08G61		318.90	1076.10	0.86	0.711761	0.000015	0.710175	3.300000	23.000000	0.086200	0.511693	0.000013	−16.60	

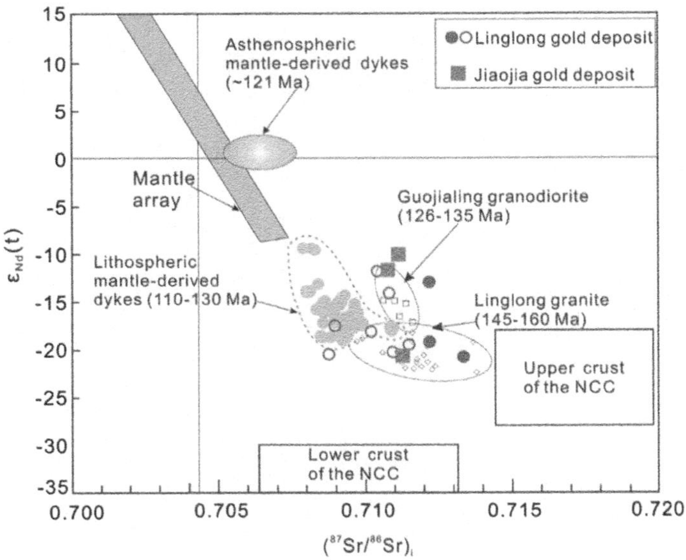

Fig. 4.14 Plots of (a) $(^{87}Sr/^{86}Sr)_i$ versus $\varepsilon_{Nd}(t)$ for the pyrites from the Linglong and Jiaojia gold deposits. *Data sources* pyrites in Linglong gold deposit (Yang and Zhou 2001), Linglong granite (Yang et al. 2012; Ma et al. 2013), Gojialing granodiorite (Zhang et al. 2010; Yang et al. 2012), Cretaceous lithospheric mantle-derived mafic dykes (Liu et al. 2009; Ma et al. 2014; Deng et al. 2017; Liang et al. 2019), and asthenospheric mantle-derived mafic dykes (Ma et al. 2014). Upper and lower continental crust information for the NCC is from Jahn et al. (1999)

The D–O isotopes of the Jiaojia-type gold deposits in the Jiaodong Peninsula were plotted outside the metamorphic water field due to their relatively low $\delta^{18}O$ values. These values were distributed to the primary magmatic water area, with a few points near the meteoric water line (Fig. 4.16), thereby suggesting that the ore-forming fluids were mainly sourced from primary magmatic water and a smaller addition of meteoric water. The D-O isotopes of the Linglong-type gold deposits were mainly distributed in the lower left part of the metamorphic water field and the primary magmatic water area (Fig. 4.16), thereby indicating that the magmatic water was the main source of the gold deposits; however, mixing with meteoric waters was only documented in the Jiaojia-type gold deposit. Deng et al. (2019) proved that the regional peak metamorphism of the Jiaodong Peninsula occurred at 1.9–2.2 Ga and a craton-margin orogeny occurred at ~230 Ma, both of which were older than the period of mineralisation. Wen et al. (2015) showed that the age of mineralisation was also younger than the ages of the Linglong (160–156 Ma) and Guojialing (130–126 Ma) plutons in the Jiaobei terrane. However, late Mesozoic sub-continental lithospheric mantle (SCLM)-derived mafic dykes (110–130 Ma; Deng et al. 2017) emplaced at Jiaodong Peninsula have demonstrated a close temporal and spatial relationship with gold mineralisation. Thus, mantle-derived magmatic water is likely the main source of the ore-forming fluid, suggesting that mantle-derived

Table 4.7 Sulfur isotopes of pyrite in the Linglong and Jiaojia gold deposits

Sample locations	Measured minerals	$\delta^{34}S$‰		
		Stage I	Stage II	Stage III
Linglong	Pyrite	7.65	7.9	9.09
	Pyrite	7.98	7.44	8.13
	Pyrite	8.3	7.01	8.29
	Pyrite	8.1	7.64	8.54
	Pyrite	8.35	6.67	8.78
	Pyrite	8.86	7.12	9.16
	Pyrite	8.56	6.49	6.98
	Pyrite	7.55	6.27	8.05
	Pyrite	8.13	9.4	8.16
	Pyrite	8.48	9.8	9.11
	Pyrite	9.35	8.78	8.28
	Pyrite	8.8	8.62	8.15
	Pyrite	8.06	9.34	8.75
	Pyrite	7.92	9.46	9.24
	Pyrite	8.17	8.07	9.6
	Pyrite	8.42	8.17	8.54
	Pyrite	7.99	8.63	8.06
	Pyrite	7.16	8.56	8.02
	Pyrite	9.39	8.2	8.42
	Pyrite	8.43	8.35	8.63
	Pyrite	8.47	8.6	8.85
Jiaojia	Pyrite	10.84	10.64	10.7
	Pyrite	11.19	9.81	10.53
	Pyrite	10.3	10.06	10.83
	Pyrite	9.46	10.36	9.97
	Pyrite	10.52	10.82	10.76
	Pyrite	10.56	8.43	10.71
	Pyrite	9.5	10.79	8.98
	Pyrite	10.55	8.75	10.67
	Pyrite	10.23		

fluids play an important role in the metallogenic process; this finding is consistent with previous studies (Deng et al. 2003; Liu et al. 2003; Mao et al. 2005). Deng et al. (2019) also reported that the C–O isotopes of the Sanshandao gold deposit showed a mantle-like source composition.

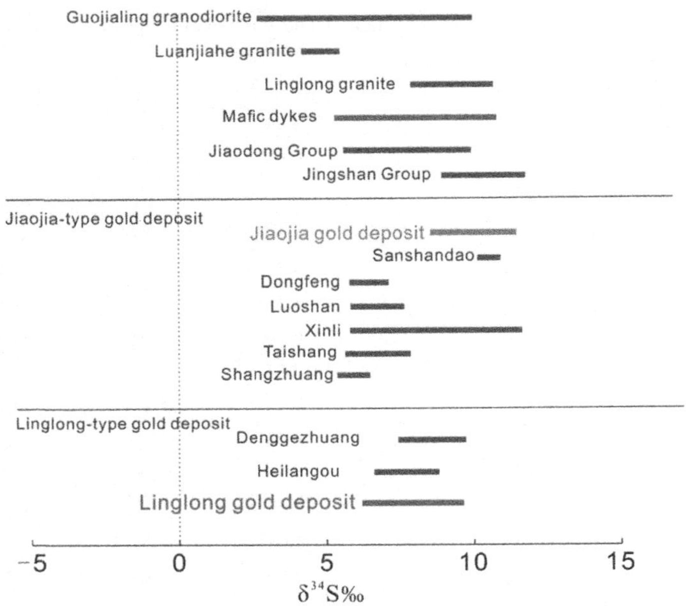

Fig. 4.15 Diagram of sulfur isotopic composition of pyrite from the Linglong and Jiaojia gold deposits and regional rocks. The sulfur isotopic data are from Mao et al. (2008), Wen et al. (2015), Yang et al. (2016b), Cai et al. (2018) and Deng et al. (2019)

Although $\delta^{34}S$ values of pyrites from the Linglong and Jiaojia gold deposits are within the $\delta^{34}S$ ranges of Late Jurassic granites, early Cretaceous granodiorites, the Archean Jiaodong Group, and the Paleoproterozoic Jingshan group in the Jiaodong Peninsula (Fig. 4.15), these sources were also excluded for occurring earlier than mineralization. The high $\delta^{34}S$ values were also likely derived from seawater mixing (Rowland and Sibson 2004) or progressive magmatic oxidation (Mills et al. 2015b). There were no Palaeozoic to late Mesozoic strata nor any geological evidence explaining the nature of the Cretaceous sea levels in the Jiaodong Peninsula (Mills et al. 2015b); therefore, seawater mixing is impossible. However, the Cretaceous lithospheric mantle-derived basaltic rocks in the eastern NCC have a high oxygen fugacity (+1.5 ~+1.9), which are obtained from the most forsteritic olivine phenocrysts ($Mg^{\#} > 89$, and up to 91.5) (Geng et al. 2019). Thus, the $\delta^{34}S$ values of pyrite also suggest mantle-derived magmatic fluids (Table 4.9).

4.4.4 *Source of Ore-Forming Materials*

Pyrites in the Linglong gold deposit showed $(^{87}Sr/^{86}Sr)_i$ values of 0.71215–0.71331 (average value of 0.713), and $\varepsilon_{Nd}(t)$ values of −20. 6 to −12.7 (average value of

Table 4.8 Sulfur isotopes of pyrite in sulfide ores and rocks from the Jiaodong gold province

Sample locations	Sample no.	Stage	Measured minerals	$\delta^{34}S_{CDT}$/‰	Data sources
Jijiao- type deposits					
Sanshandao	SS-435-1		Ser, Pyr	12	Mao et al. (2008)
Sanshandao	SS-70-13		Ser, Pyr	12	Mao et al. (2008)
Sanshandao	SS-375-6		Pyr	11.7	Mao et al. (2008)
Sanshandao	SS-375-2		Pyr	11.9	Mao et al. (2008)
Sanshandao	SSD2		Pyr	11.5	Mao et al. (2008)
Sanshandao	Csh1		Ser, Pyr	11.7	Mao et al. (2008)
Sanshandao	Csh2		Ser, Pyr	12.4	Mao et al. (2008)
Sanshandao	Csh5		Ser, Pyr	11.3	Mao et al. (2008)
Sanshandao	Csh6		Ser, Pyr	12.5	Mao et al. (2008)
Sanshandao	Csh7		Ser, Pyr	11.7	Mao et al. (2008)
Jiaojia	JJ-190-2		Ser, Pyr	12.2	Mao et al. (2008)
Jiaojia	JJ-190-13		K-feld, Pyr	11.4	Mao et al. (2008)
Jiaojia	JJ-190-14		K-feld, Pyr	11.5	Mao et al. (2008)
Jiaojia	JJ-190-10		Pyr	11.3	Mao et al. (2008)
Jiaojia	Jjia3		Pyr	10.1	Mao et al. (2008)
Dongfeng	10LL18	Middle stage	Pyrite	6.7	Wen et al. (2015)
Dongfeng	10LL20	Middle stage	Pyrite	6.5	Wen et al. (2015)
Dongfeng	10LL22	Middle stage	Pyrite	6.8	Wen et al. (2015)
Dongfeng	10LL23	Middle stage	Pyrite	5.8	Wen et al. (2015)
Dongfeng	10LL24	Middle stage	Pyrite	7	Wen et al. (2015)

(continued)

Table 4.8 (continued)

Sample locations	Sample no.	Stage	Measured minerals	$\delta^{34}S_{CDT}$/‰	Data sources
Luoshan	LS13D001B2	Early stage ore	Pyrite	6.5	Guo et al. (2017)
Luoshan	LS13D002B2	Early stage ore	Pyrite	5.8	Guo et al. (2017)
Luoshan	LS13D014B1	Early stage ore	Pyrite	6.7	Guo et al. (2017)
Luoshan	LS13D024B1	Early stage ore	Pyrite	4.3	Guo et al. (2017)
Luoshan	LS13D029B1	Early stage ore	Pyrite	6.5	Guo et al. (2017)
Luoshan	LS13D064B1	Early stage ore	Pyrite	6.9	Guo et al. (2017)
Luoshan	LS13D080B1	Early stage ore	Pyrite	6	Guo et al. (2017)
Luoshan	LS13D002B1	Early stage ore	Pyrite	7.5	Guo et al. (2017)
Luoshan	LS13D008B1	Main stage ore	Pyrite	7	Guo et al. (2017)
Luoshan	LS13D014B4	Main stage ore	Pyrite	6.7	Guo et al. (2017)
Luoshan	LS13D014B6	Main stage ore	Pyrite	7.2	Guo et al. (2017)
Luoshan	LS13D036B3	Main stage ore	Pyrite	7.5	Guo et al. (2017)
Luoshan	LS13D036B5	Main stage ore	Pyrite	7.2	Guo et al. (2017)
Luoshan	LS13D036B6	Main stage ore	Pyrite	7.5	Guo et al. (2017)
Luoshan	LS13D038B1	Main stage ore	Pyrite	7.3	Guo et al. (2017)
Luoshan	LS13D043B2	Main stage ore	Pyrite	7.5	Guo et al. (2017)
Luoshan	LS13D048B2	Main stage ore	Pyrite	7	Guo et al. (2017)
Luoshan	LS13D049B2	Main stage ore	Pyrite	7.4	Guo et al. (2017)
Luoshan	LS13D051B1	Main stage ore	Pyrite	7.2	Guo et al. (2017)
Luoshan	LS13D063B1	Main stage ore	Pyrite	5.5	Guo et al. (2017)
Luoshan	LS13D066B1	Main stage ore	Pyrite	7.3	Guo et al. (2017)

(continued)

Table 4.8 (continued)

Sample locations	Sample no.	Stage	Measured minerals	$\delta^{34}S_{CDT}$/‰	Data sources
Luoshan	LS13D067B1	Main stage ore	Pyrite	6.6	Guo et al. (2017)
Luoshan	LS13D067B2	Main stage ore	Pyrite	6.2	Guo et al. (2017)
Luoshan	LS13D068B1	Main stage ore	Pyrite	6.7	Guo et al. (2017)
Luoshan	LS13D071B2	Main stage ore	Pyrite	7.1	Guo et al. (2017)
Luoshan	LS13D073B1	Main stage ore	Pyrite	7.4	Guo et al. (2017)
Luoshan	FS13D01B2	Main stage ore	Pyrite	5.8	Guo et al. (2017)
Luoshan	FS13D31B2	Main stage ore	Pyrite	7.1	Guo et al. (2017)
Luoshan	FS13D33B3	Main stage ore	Pyrite	6.8	Guo et al. (2017)
Xinli gold	X0608–15	1st	Pyrite	9.54	Deng et al. (2015a)
Xinli gold	X0608–24	1st	Pyrite	10.34	Deng et al. (2015a)
Xinli gold	X0635–28	1st	Pyrite	9.75	Deng et al. (2015a)
Xinli gold	X0635–30	1st	Pyrite	10.5	Deng et al. (2015a)
Xinli gold	X0635–32	1st	Pyrite	10.11	Deng et al. (2015a)
Xinli gold	X0635–33	1st	Pyrite	10.03	Deng et al. (2015a)
Xinli gold	X0635–41	1st	Pyrite	10.18	Deng et al. (2015a)
Xinli gold	X0506–21–1	2nd	Pyrite	9.42	Deng et al. (2015a)
Xinli gold	X0506–26	2nd	Pyrite	9.9	Deng et al. (2015a)
Xinli gold	X0506–29	2nd	Pyrite	10.16	Deng et al. (2015a)
Xinli gold	X0835–44	3rd	Pyrite	10.72	Deng et al. (2015a)
Xinli gold	X0835–65	3rd	Pyrite	9.99	Deng et al. (2015a)
Xinli gold	X0904–23	3rd	Pyrite	11.62	Deng et al. (2015a)

(continued)

Table 4.8 (continued)

Sample locations	Sample no.	Stage	Measured minerals	$\delta^{34}S_{CDT}$/‰	Data sources
Xinli gold	X0937–10	3rd	Pyrite	10.44	Deng et al. (2015a)
Xinli gold	X1035–5	3rd	Pyrite	10.43	Deng et al. (2015a)
Xinli gold	X0608–10	3rd	Galena	9.83	Deng et al. (2015a)
Xinli gold	X0608–10–1	3rd	Galena	5.81	Deng et al. (2015a)
Xinli gold	X0835–44–1	3rd	Galena	6.81	Deng et al. (2015a)
Xinli gold	X0635–15	3th	Galena	7.02	Deng et al. (2015a)
Xinli gold	X0635–35	3th	Galena	8.27	Deng et al. (2015a)
Xinli gold	X0506–22	3th	Sphalerite	9.64	Deng et al. (2015a)
Xinli gold	X0635–15–1	3th	Sphalerite	9.87	Deng et al. (2015a)
Xinli gold	X0635–35–1	3th	Sphalerite	10.49	Deng et al. (2015a)
Xinli gold	X0904–16–1	3th	Sphalerite	9.41	Deng et al. (2015a)
Xinli gold	X0904–16–2	3th	Sphalerite	9.55	Deng et al. (2015a)
Taishang	Ts12D023B3	Stage 1	Pyrite	7.2	Yang et al. (2016a)
Taishang	Ts12D025B4	Stage 1	Pyrite	6.8	Yang et al. (2016a)
Taishang	Ts12D028B5	Stage 1	Pyrite	7.4	Yang et al. (2016a)
Taishang	Ts12D033B2	Stage 1	Pyrite	6.3	Yang et al. (2016a)
Taishang	Ts12D055B4	Stage 1	Pyrite	4.5	Yang et al. (2016a)
Taishang	Ts12D016B2	Stage 1	Pyrite	6.3	Yang et al. (2016a)
Taishang	Ts12D006B1	Stage 2	Pyrite	5.7	Yang et al. (2016a)
Taishang	Ts12D018B1	Stage 2	Pyrite	6.6	Yang et al. (2016a)

(continued)

Table 4.8 (continued)

Sample locations	Sample no.	Stage	Measured minerals	$\delta^{34}S_{CDT}$/‰	Data sources
Taishang	Ts12D023B1	Stage 2	Pyrite	7.8	Yang et al. (2016a)
Taishang	Ts12D023B4	Stage 2	Pyrite	8	Yang et al. (2016a)
Taishang	Ts12D028B2	Stage 2	Pyrite	7.3	Yang et al. (2016a)
Taishang	Ts12D035B2	Stage 2	Pyrite	7.2	Yang et al. (2016a)
Taishang	Ts12D045B1	Stage 2	Pyrite	6.4	Yang et al. (2016a)
Taishang	Ts12D045B2	Stage 2	Pyrite	7.3	Yang et al. (2016a)
Taishang	Ts12D055B2	Stage 2	Pyrite	6.2	Yang et al. (2016a)
Taishang	Ts12D003B1	Stage 2	Pyrite	6.8	Yang et al. (2016a)
Taishang	Ts12D012B2	Stage 2	Pyrite	7.3	Yang et al. (2016a)
Taishang	Ts12D017B2	Stage 2	Pyrite	5.9	Yang et al. (2016a)
Taishang	Ts12D024B1	Stage 2	Pyrite	6.3	Yang et al. (2016a)
Taishang	Ts12D023B2	Stage 3	Pyrite	7.5	Yang et al. (2016a)
Taishang	Ts12D045B3	Stage 3	Pyrite	7.9	Yang et al. (2016a)
Taishang	Ts12D012B1	Stage 3	Pyrite	6.4	Yang et al. (2016a)
Taishang	Ts12D034B1	Stage 3	Pyrite	7.1	Yang et al. (2016a)
Taishang	Ts12D043B3	Stage 3	Pyrite	7.9	Yang et al. (2016a)
Taishang	Ts12D010B1	Stage 3	Pyrite	6.8	Yang et al. (2016a)
Shangzhuang	13SZ47-1	II	Pyrite	6.5	Cai et al. (2018)
Shangzhuang	13SZ47-2	II	Pyrite	4.8	Cai et al. (2018)
Shangzhuang	13SZ47-4	II	Pyrite	4.6	Cai et al. (2018)

(continued)

Table 4.8 (continued)

Sample locations	Sample no.	Stage	Measured minerals	$\delta^{34}S_{CDT}$/‰	Data sources
Shangzhuang	13SZ47-5	II	Pyrite	5.6	Cai et al. (2018)
Shangzhuang	13SZ47-8	II	Pyrite	6.3	Cai et al. (2018)
Shangzhuang	13SZ47-9	II	Pyrite	5.7	Cai et al. (2018)
Shangzhuang	13SZ47-2	II	Molybdenite	5.5	Cai et al. (2018)
Shangzhuang	13SZ47-4	II	Molybdenite	5.3	Cai et al. (2018)
Linglong-type ore					
Dayigezhuang	Dygz4		Ser, Pyr	7.9	Mao et al. (2008)
Dayigezhuang	Dygz5		Ser		Mao et al. (2008)
Dayigezhuang	Dygz7		Ser, Pyr	7.5	Mao et al. (2008)
Dayigezhuang	Dygz8		Ser, Pyr	7.5	Mao et al. (2008)
Dayigezhuang	Dygz9		Ser, Pyr	7.2	Mao et al. (2008)
Dayigezhuang	Dygz1		K-feld		Mao et al. (2008)
Dayigezhuang	Dygz6		K-feld, Pyr	7.8	Mao et al. (2008)
Dayigezhuang	Dygz10		K-feld, Pyr	6.8	Mao et al. (2008)
Dayigezhuang	Dygz3		Pyr	7.6	Mao et al. (2008)
Denggezhuang	Dgz2		Ser, Pyr	9.3	Mao et al. (2008)
Denggezhuang	Dgz11		Ser, Pyr	9	Mao et al. (2008)
Denggezhuang	Dgz5		K-feld		Mao et al. (2008)
Denggezhuang	Dgz10		K-feld		Mao et al. (2008)
Denggezhuang	Dgz3		Pyr	9.6	Mao et al. (2008)

(continued)

Table 4.8 (continued)

Sample locations	Sample no.	Stage	Measured minerals	$\delta^{34}S_{CDT}$/‰	Data sources
Denggezhuang	Dgz4		Pyr	9.3	Mao et al. (2008)
Denggezhuang	Dgz6		Pyr	9.3	Mao et al. (2008)
Denggezhuang	Dgz7		Pyr	8.8	Mao et al. (2008)
Linglong	10LL03	Early	Pyrite	6.6	Wen et al. (2015)
Linglong	JQ-Q-04	Middle	Pyrite	6.5	Wen et al. (2015)
Linglong	LL-Q-02	Middle	Pyrite	6.2	Wen et al. (2015)
Linglong	LL-Q-06	Middle	Pyrite	5.9	Wen et al. (2015)
Linglong	10X80	Late	Pyrite	7.4	Wen et al. (2015)
Linglong	LL-108-1	Middle	Pyrite	7.7	Hou et al. (2006)
Linglong	LL-108-2	Middle	Pyrite	8.3	Hou et al. (2006)
Linglong	LL-108-5	Middle	Pyrite	7.6	Hou et al. (2006)
Linglong	LL-108-6	Middle	Pyrite	8.5	Hou et al. (2006)
Linglong	LL-108-7	Middle	Pyrite	7.5	Hou et al. (2006)
Linglong	LL-53-4	Middle	Pyrite	7	Hou et al. (2006)
Linglong	LL-48-1	Middle	Pyrite	8.6	Hou et al. (2006)
Linglong	LL-48-4	Middle	Pyrite	7.2	Hou et al. (2006)
Linglong	LL-50-3	Middle	Pyrite	6.4	Hou et al. (2006)
Linglong	LL-50-5	Middle	Pyrite	7.3	Hou et al. (2006)
Linglong	LL-47-1	Middle	Pyrite	7.4	Hou et al. (2006)
Lingnan	LL-171-2	Middle	Pyrite	7.8	Hou et al. (2006)

(continued)

Table 4.8 (continued)

Sample locations	Sample no.	Stage	Measured minerals	$\delta^{34}S_{CDT}$/‰	Data sources
Lingnan	LL-171-3	Middle	Pyrite	7	Hou et al. (2006)
Lingnan	LL-171-6	Middle	Pyrite	8.3	Hou et al. (2006)
Lingnan	LL-171-7	Middle	Pyrite	7.8	Hou et al. (2006)
Lingnan	LL-171-10	Middle	Pyrite	7.9	Hou et al. (2006)
Heilangou	14HLG12@2	Py-I	Pyrite	6.8	Feng et al. (2018)
Heilangou	14HLG12@3	Py-I	Pyrite	6.6	Feng et al. (2018)
Heilangou	14HLG12@4	Py-I	Pyrite	6.9	Feng et al. (2018)
Heilangou	14HLG12@5	Py-I	Pyrite	7.1	Feng et al. (2018)
Heilangou	14HLG12@6	Py-I	Pyrite	7.1	Feng et al. (2018)
Heilangou	15HLG54-3@1	Py-I	Pyrite	7	Feng et al. (2018)
Heilangou	15HLG54-3@2	Py-I	Pyrite	7	Feng et al. (2018)
Heilangou	15HLG54-3@3	Py-I	Pyrite	7.2	Feng et al. (2018)
Heilangou	15HLG54-3@4	Py-I	Pyrite	7	Feng et al. (2018)
Heilangou	15HLG54-3@5	Py-I	Pyrite	7	Feng et al. (2018)
Heilangou	15HLG54-3@6	Py-I	Pyrite	7.1	Feng et al. (2018)
Heilangou	15HLG54-3@7	Py-I	Pyrite	7.2	Feng et al. (2018)
Heilangou	14HLG24R@1	Py-I	Pyrite	7.9	Feng et al. (2018)
Heilangou	14HLG24R@2	Py-I	Pyrite	7.2	Feng et al. (2018)
Heilangou	14HLG24R@3	Py-I	Pyrite	7.2	Feng et al. (2018)
Heilangou	14HLG24R@4	Py-I	Pyrite	7.5	Feng et al. (2018)

(continued)

Table 4.8 (continued)

Sample locations	Sample no.	Stage	Measured minerals	$\delta^{34}S_{CDT}$/‰	Data sources
Heilangou	14HLG24R@5	Py-I	Pyrite	7.7	Feng et al. (2018)
Heilangou	14HLG24R@6	Py-I	Pyrite	7.6	Feng et al. (2018)
Heilangou	14HLG14-2@1	Py-IIb	Pyrite	8.5	Feng et al. (2018)
Heilangou	14HLG14-2@2	Py-IIb	Pyrite	8.4	Feng et al. (2018)
Heilangou	14HLG14-2@3	Py-IIb	Pyrite	8.3	Feng et al. (2018)
Heilangou	14HLG14-2@4	Py-IIb	Pyrite	8.3	Feng et al. (2018)
Heilangou	14HLG14-2@5	Py-IIa	Pyrite	7.7	Feng et al. (2018)
Heilangou	14HLG14-2@6	Py-IIa	Pyrite	7.8	Feng et al. (2018)
Heilangou	14HLG14-2@8	Py-IIa	Pyrite	7.5	Feng et al. (2018)
Heilangou	14HLG14-2@9	Py-IIb	Pyrite	8.5	Feng et al. (2018)
Heilangou	14HLG14-2@10	Py-IIb	Pyrite	8.5	Feng et al. (2018)
Heilangou	14HLG14-2@11	Py-IIb	Pyrite	8.4	Feng et al. (2018)
Heilangou	14HLG14-2@12	Py-IIb	Pyrite	8.3	Feng et al. (2018)
Heilangou	14HLG14-2@13	Py-IIa	Pyrite	7.7	Feng et al. (2018)
Heilangou	14HLG14-2@14	Py-IIa	Pyrite	7.4	Feng et al. (2018)
Heilangou	14HLG14-2@15	Py-IIb	Pyrite	8.4	Feng et al. (2018)
Heilangou	14HLG14-2@16	Py-IIb	Pyrite	8.3	Feng et al. (2018)
Heilangou	14HLG14-2@17	Py-IIb	Pyrite	8.3	Feng et al. (2018)
Heilangou	14HLG14-2@18	Py-IIa	Pyrite	7.7	Feng et al. (2018)
Heilangou	14HLG14-2@19	Py-IIb	Pyrite	8.6	Feng et al. (2018)

(continued)

Table 4.8 (continued)

Sample locations	Sample no.	Stage	Measured minerals	$\delta^{34}S_{CDT}$/‰	Data sources
Heilangou	14HLG14-2@20	Py-IIb	Pyrite	8.8	Feng et al. (2018)
Heilangou	14HLG14@1	Py-IIb	Pyrite	8.4	Feng et al. (2018)
Heilangou	14HLG14@2	Py-IIb	Pyrite	8.6	Feng et al. (2018)
Heilangou	14HLG14@5	Py-IIa	Pyrite	7.3	Feng et al. (2018)
Heilangou	14HLG14@6	Py-IIa	Pyrite	7.7	Feng et al. (2018)
Heilangou	14HLG14@7	Py-IIb	Pyrite	8.3	Feng et al. (2018)
Heilangou	14HLG14@8	Py-IIb	Pyrite	8.1	Feng et al. (2018)
Heilangou	14HLG14@9	Py-IIb	Pyrite	8.4	Feng et al. (2018)
Heilangou	14HLG13R@1	Py-IIa	Pyrite	7.1	Feng et al. (2018)
Heilangou	14HLG13R@2	Py-IIa	Pyrite	6.9	Feng et al. (2018)
Heilangou	14HLG13R@3	Py-IIa	Pyrite	7.6	Feng et al. (2018)
Heilangou	14HLG13R@4	Py-IIa	Pyrite	7.8	Feng et al. (2018)
Heilangou	14HLG13R@5	Py-IIb	Pyrite	8.4	Feng et al. (2018)
Heilangou	14HLG13R@6	Py-IIb	Pyrite	8.2	Feng et al. (2018)
Heilangou	14HLG13R@7	Py-IIb	Pyrite	8.2	Feng et al. (2018)
Heilangou	14HLG13R@8	Py-IIb	Pyrite	8.3	Feng et al. (2018)
Heilangou	14HLG13R@9	Py-IIb	Pyrite	8.7	Feng et al. (2018)
Heilangou	14HLG13R@10	Py-IIa	Pyrite	7.7	Feng et al. (2018)
Heilangou	14HLG13R@11	Py-IIa	Pyrite	7.7	Feng et al. (2018)
Heilangou	14HLG13R@12	Py-IIb	Pyrite	8.2	Feng et al. (2018)

(continued)

Table 4.8 (continued)

Sample locations	Sample no.	Stage	Measured minerals	$\delta^{34}S_{CDT}$/‰	Data sources
Heilangou	14HLG13R@13	Py-IIb	Pyrite	8.7	Feng et al. (2018)
Heilangou	14HLG13R@14	Py-IIb	Pyrite	8.4	Feng et al. (2018)
Heilangou	14HLG13R@15	Py-IIb	Pyrite	8.5	Feng et al. (2018)
Heilangou	15HLG54-1@1	Py-IIb	Pyrite	8.3	Feng et al. (2018)
Heilangou	15HLG54-1@2	Py-IIb	Pyrite	8.4	Feng et al. (2018)
Heilangou	15HLG54-1@3	Py-IIa	Pyrite	7.2	Feng et al. (2018)
Heilangou	15HLG54-1@4	Py-IIa	Pyrite	7.1	Feng et al. (2018)
Heilangou	15HLG54-1@5	Py-IIa	Pyrite	7.4	Feng et al. (2018)
Heilangou	15HLG54-1@6	Py-IIb	Pyrite	8.5	Feng et al. (2018)
Granitoids					
Linglong	LL-190-11		Pyr	7.9	Mao et al. (2008)
Linglong	LL-190-1		K-feld	8.2	Mao et al. (2008)
Linglong	LL-190-47-4		Pyr	8.3	Mao et al. (2008)
Linglong	LL-190-5		Pyr	8.4	Mao et al. (2008)
Linglong	LL-190-6		Pyr	8.2	Mao et al. (2008)
Linglong	LL-190-7		Pyr	8	Mao et al. (2008)
Wang'ershan	WE-270-V-8		Pyr	8.6	Mao et al. (2008)

Bio—Biotite; K-feld—K-feldspar; Min—Mineral; Mus—Muscovite; Pyr—Pyrite; Ser—Sericite

−17.4). Pyrites in the Jiaojia gold deposit showed $(^{87}Sr/^{86}Sr)_i$ values of 0.71075–0.71125 (average value of 0.711) and ε_{Nd} (t) values of −20.5 to −9.83 (average value of −13.9) (Table 4.3). These Sr-Nd isotope compositions of gold deposits differed from those of the lower and upper continental crust of NCC (Fig. 4.17), excluding continental crust sources. Moreover, in the continental crust-normalised diagram of Fig. 4.13a, the Linglong and Jiaojia gold deposits show significant enrichment of

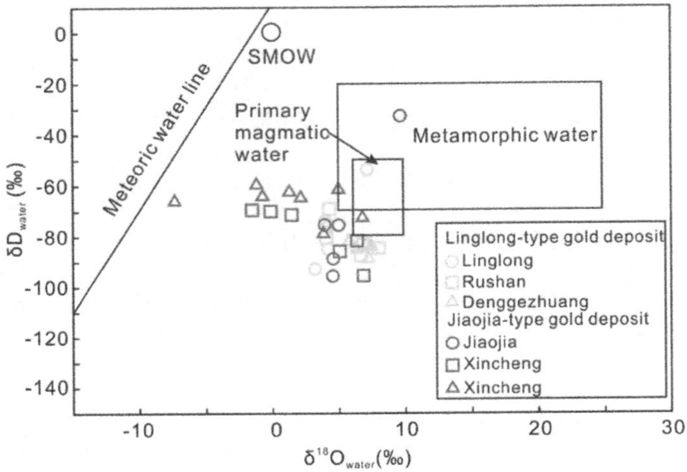

Fig. 4.16 Isotopic compositions of oxygen and hydrogen from the Linglong-type and Jiaojia-type gold deposits. The D-O isotopic data are from Mao et al. (2008), Wen et al. (2015), Guo et al. (2017), Deng et al. (2015a), Yang et al. (2016b), and Cai et al. (2018). The isotopic fields for common geological waters are from Hugh and Taylor (1974)

chalcophile elements (Cu, Pb, Zn, and Cd), which could also exclude crustal sources or crustal contamination. Although pyrites in the Linglong and Jiaojia gold deposits showed similar Sr-Nd isotopic compositions to Late Jurassic Linglong granite (160–156 Ma) and Early Cretaceous Guojialing granodiorite (130–126 Ma) (Fig. 4.17), the ages of these granites were older than the gold mineralisation age, which also removed the possibility of a granitic magmatic source. Moreover, the Linglong and Jiaojia gold deposits had lower $\varepsilon_{Nd}(t)$ values than those of the Archean Jiaodong Group and the Paleoproterozoic Jingshan Group in the Jiaodong Peninsula (Fig. 4.17). The high-grade metamorphic ancient crust could be highly depleted in gold and fluids (Goldfarb and Santosh 2014; Goldfarb and Groves 2015; Wang et al. 2020). Thus, the Precambrian metamorphic basement was also not the source of ore-forming materials. However, the Jiaobei Cretaceous SCLM-derived mafic dykes (110–130 Ma) have a limited range of $(^{87}Sr/^{86}Sr)_i$ isotope ratios (0.709–0.711) and $\varepsilon_{Nd}(t)$ values (−18.5 to −13.9) (Ma et al. 2014; Deng et al. 2017), which is close to the pyrites of Linglong and Jiaojia gold deposits (Fig. 4.17). Previous studies have shown that SCLM underneath the Archean crust is refractory and strongly depleted in incompatible elements (large ion lithophile elements, heavy rare earth elements, HFSEs, V, Ga, and Cr) and many metals (e.g., Au and Platinum-group elements). However, the metasomatised SCLM is enriched in Au (Griffn et al. 2009; Hronsky et al. 2012; Lorand et al. 2013; Tassara et al. 2017; Wang et al. 2019a, b; Qiu et al. 2019; Zhao et al. 2019). Wang et al. (2020) studied the Au contents of the metasomatised mantle xenoliths (480-4 Ma; 0.02–1.8 ppb Au), metasomatised SCLM-derived basalts (130–120 Ma; 2–3 ppb Au) and asthenosphere-derived basalts (<110 Ma; 0.04–1.26 ppb Au), and found that although primitive basaltic magmas have low Au contents, the

Table 4.9 Hydrogen and oxygen isotope compositions from gold deposit in Jiaodong Peninsula

Sample locations	Sample no.	Stage	Measured minerals	δD (‰)SMOW	$\delta^{18}O_{mineral}$ (‰)SMOW	$\delta^{18}O_{water}$ (‰)SMOW	Data sources
Jiaojia-type deposits							
Sanshandao	SS-435-1		Ser	−52	10.7	7.55	Mao et al. (2008)
Sanshandao	SS-70-13		Ser	−48	11.7	8.55	Mao et al. (2008)
Sanshandao	SS-375-4		Ser	−52	12	8.85	Mao et al. (2008)
Sanshandao	SS-375-7		Ser	−53	9.7	6.55	Mao et al. (2008)
Sanshandao	SSD1		Ser	−50	10.9	7.75	Mao et al. (2008)
Sanshandao	SSD3		Ser	−67	11	7.85	Mao et al. (2008)
Sanshandao	Csh1		Ser	−56	10.9	7.75	Mao et al. (2008)
Sanshandao	Csh2		Ser	−52	10	6.85	Mao et al. (2008)
Sanshandao	Csh5		Ser	−51	12.8	9.65	Mao et al. (2008)
Sanshandao	Csh6		Ser	−48	11.4	8.25	Mao et al. (2008)
Sanshandao	Csh7		Ser	−52	10.5	7.35	Mao et al. (2008)
Jiaojia	JJ-190-2		Ser	−62	10.4	7.25	Mao et al. (2008)
Jiaojia	Jjia1		Ser	−59	10.9	7.75	Mao et al. (2008)
Jiaojia	Jjia2		Ser	−59	8.4	5.25	Mao et al. (2008)
Jiaojia	JJ-190-13		K-feld	−86	10.2	4.37	Mao et al. (2008)
Jiaojia	JJ-190-14		K-feld	−85	10	4.17	Mao et al. (2008)
Wang’ershan	WE-270-V-3		Ser	−57	9.1	5.95	Mao et al. (2008)
Wang’ershan	WE-310-V-12		Ser	−52	8.7	5.55	Mao et al. (2008)
Wang’ershan	WE-130-2		Ser	−59	10.4	7.25	Mao et al. (2008)
Wang’ershan	WE-270-V-5		Ser	−61	10.1	6.95	Mao et al. (2008)

(continued)

Table 4.9 (continued)

Sample locations	Sample no.	Stage	Measured minerals	δD (‰)SMOW	$\delta^{18}O_{mineral}$ (‰)SMOW	$\delta^{18}O_{water}$ (‰)SMOW	Data sources
Dongfeng	PZ46812-1	Early	Quartz	−82.7	6.3	−0.8	Wen et al. (2015)
Dongfeng	PZ4966-1	Middle	Quartz	−85.7	6.4	−1.7	Wen et al. (2015)
Dongfeng	PZ41207-2	Middle	Quartz	−90.5	14.5	6.4	Wen et al. (2015)
Dongfeng	PZ46812-3	Late	Quartz	−87.5	11.6	−3.8	Wen et al. (2015)
Luoshan	LS13D036B1		Quartz	−68	13.8	8	Guo et al. (2017)
Luoshan	LS13D002B1		Quartz	−53	9.1	3.1	Guo et al. (2017)
Luoshan	LS13D038B1		Quartz	−45	6.3	0.3	Guo et al. (2017)
Luoshan	LS13D043B2		Quartz	−40	7.9	1.9	Guo et al. (2017)
Xinli	X0506–26	2nd	Quartz	−74.4	13.8	5.34	Deng et al. (2015a)
Xinli	X0635–9	2nd	Quartz	−79.5	12 4.	2	Deng et al. (2015a)
Xinli	X0635–12	2nd	Quartz	−73.7	10.3	3.65	Deng et al. (2015a)
Xinli	X0635–41	2nd	Quartz	−84.5	11.9	3.92	Deng et al. (2015a)
Xinli	X0635–52	2nd	Quartz	−88.3	12.9	3.21	Deng et al. (2015a)
Xinli	X0835–69	2nd	Quartz	−79.8	12.1	3.79	Deng et al. (2015a)
Xinli	X0835–44	3rd	Quartz	−70.8	12 3	0.03	Deng et al. (2015a)
Xinli	X1035–5	3rd	Quartz	−69.6	12.1	2.82	Deng et al. (2015a)
Taishang	Ts12D025B4	Stage 3	Quartz	−77	15.6	9.9	Yang et al. (2016b)
Taishang	Ts12D028B5	Stage 4	Quartz	−78	11.4	5.7	Yang et al. (2016b)
Taishang	Ts12D055B4	Stage 7	Quartz	−80	8.5	2.8	Yang et al. (2016b)
Taishang	Ts12D042B1	Stage 11	Quartz	−76	12.5	6.8	Yang et al. (2016b)
Taishang	Ts12D018B1	Stage 3	Quartz	−87			Yang et al. (2016b)

(continued)

Table 4.9 (continued)

Sample locations	Sample no.	Stage	Measured minerals	δD (‰)SMOW	$\delta^{18}O_{mineral}$ (‰)SMOW	$\delta^{18}O_{water}$ (‰)SMOW	Data sources
Taishang	Ts12D023B4	Stage 5	Quartz	−80	13.8	5.9	Yang et al. (2016b)
Taishang	Ts12D045B2	Stage 9	Quartz	−84			Yang et al. (2016b)
Taishang	Ts12D055B2	Stage 10	Quartz	−94	9.2	1.3	Yang et al. (2016b)
Shangzhuang	13SZ02	II	Quartz	−75.2	13.7	7.3	Cai et al. (2018)
Shangzhuang	13SZ04	II	Quartz	−65.3	12.3	5.9	Cai et al. (2018)
Shangzhuang	13SZ05	II	Quartz	−70	12.9	6.5	Cai et al. (2018)
Shangzhuang	13SZ08	II	Quartz	−70.8	12.9	6.5	Cai et al. (2018)
Shangzhuang	13SZ47-1	III	Quartz	−70.6	11.7	3.4	Cai et al. (2018)
Shangzhuang	13SZ47-9	III	Quartz	−74.7	12	3.7	Cai et al. (2018)
Linglong-type ore							
Dayigezhuang	Dygz4		Ser	−64	9.3	6.15	Mao et al. (2008)
Dayigezhuang	Dygz5		Ser	−58	9.5	6.35	Mao et al. (2008)
Dayigezhuang	Dygz7		Ser	−52	8.8	5.65	Mao et al. (2008)
Dayigezhuang	Dygz8		Ser	−58	9.2	6.05	Mao et al. (2008)
Dayigezhuang	Dygz9		Ser	−60	9.3	6.15	Mao et al. (2008)
Dayigezhuang	Dygz1		K-feld	−62	8.9	3.07	Mao et al. (2008)
Dayigezhuang	Dygz6		K-feld	−85	9.1	3.27	Mao et al. (2008)
Dayigezhuang	Dygz10		K-feld	−87	8.6	2.77	Mao et al. (2008)
Denggezhuang	Dgz2		Ser	−76	8.7	5.55	Mao et al. (2008)
Denggezhuang	Dgz11		Ser	−57	10.5	7.35	Mao et al. (2008)
Denggezhuang	Dgz5		K-feld	−75	9.6	3.77	Mao et al. (2008)

(continued)

Table 4.9 (continued)

Sample locations	Sample no.	Stage	Measured minerals	δD (‰)SMOW	$\delta^{18}O_{mineral}$ (‰)SMOW	$\delta^{18}O_{water}$ (‰)SMOW	Data sources
Denggezhuang	Dgz10		K-feld	−85	7.9	2.07	Mao et al. (2008)
Linglong	10X74	Early	Quartz	−67.1	12.5	4.9	Wen et al. (2015)
Linglong	10X78	Early	Quartz	−63.7	11.8	4.2	Wen et al. (2015)
Linglong	10LL04	Early	Quartz	−70.1	12.9	5.3	Wen et al. (2015)
Linglong	10X79	Middle	Quartz	−69.9	14.4	6	Wen et al. (2015)
Linglong	LL-Q-06	Middle	Quartz	−77.4	17.3	8.9	Wen et al. (2015)
Linglong	10X80	Late	Quartz	−66.5	14.7	0	Wen et al. (2015)

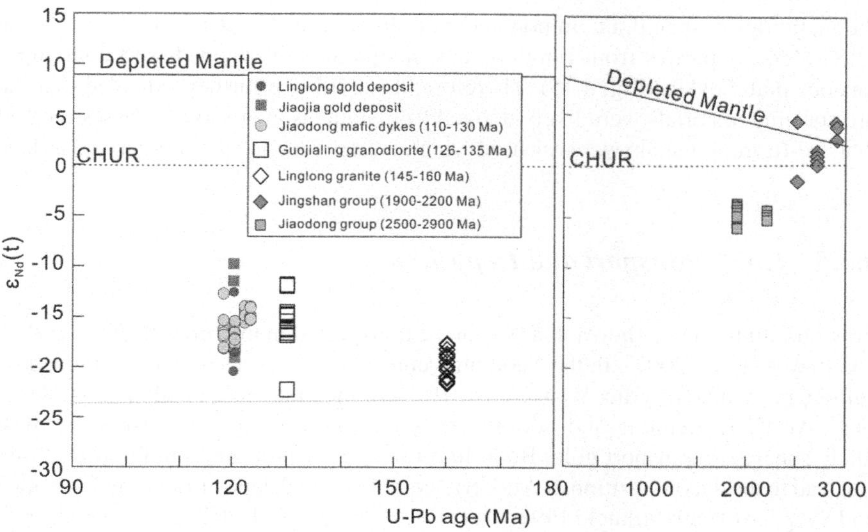

Fig. 4.17 Plots of age versus $\varepsilon_{Nd}(t)$ for pyrites from the Linglong and Jiaojia gold deposits. *Data sources* Linglong granite (Yang et al. 2012; Ma et al. 2013); Gojialing granodiorite (Zhang et al. 2010; Yang et al. 2012), Jiaobei mafic dykes (Liu et al. 2009; Ma et al. 2014; Deng et al. 2017; Liang et al. 2019), the Archean Jiaodong group (Wu et al. 2014), and the Paleoproterozoic Jingshan group (Lan et al. 2015; Li et al. 2017)

metasomatic components (high H_2O contents, S, C, and other volatiles, and elevated oxygen fugacity) in the metasomatised SCLM-derived basalts (130–120 Ma) were probably essential in producing later auriferous fluids that were exsolved and led to giant Au deposits. The lithospheric mantle beneath the eastern NCC has been recognized to be hot (~1,290–1,350 °C), water-rich (>1,000 ppm H_2O) and oxidized (log $f\,o_{2\ FQM} + 1.5 \sim +1.9$) during Early Cretaceous (Xia et al. 2013; Geng et al. 2019). Moreover, several studies have shown that the SCLM in the eastern NCC was metasomatised through subduction-related materials that may have played a key role in the large-scale Au mineralisation (Li et al. 2012; Zhu et al. 2015; Deng et al. 2019). In the Jiaodong Peninsula, 110–130 Ma mafic dykes were derived from a lithospheric mantle that was metasomatised by subduction-related fluids/melts (Guo et al. 2014; Ma et al. 2014; Cai et al. 2015; Deng et al. 2017; Liang et al. 2018). These Jiaodong mafic dykes also contained high H_2O contents of 1.38–4.45 wt% (Liang et al. 2019) and showed Pb enrichment (Ma et al. 2014; Deng et al. 2017) that was consistent with Linglong and Jiaojia pyrite. A previous LA-ICP-MS analysis of magmatic sulfide inclusions in the phenocrysts of dolerite dykes (~120 Ma) showed that metal abundance ratios (Ag, As, Au, Bi, Co, Cu, Mo, Ni, Pb, Sb, and Zn) largely corresponded to those of both rock and gold ore in the Guocheng gold deposit of Jiaodong gold province (Tan et al. 2012), which was also consistent with the enrichment of Cu, Pb, and Zn in the Linglong and Jiaojia gold deposits. In addition, the Pb isotope ratios of these mafic dykes and gold ores also demonstrated that gold mineralisation and the mafic dykes shared a common source (Tan et al. 2012, 2015). In addition, the $\delta^{34}S$ values of pyrites from Linglong and Jiaojia also overlapped with contemporaneous mafic dykes (Fig. 4.15). Therefore, this evidence further indicated that the ore-forming materials were likely derived from a metasomatised SCLM source and that ore-forming metals were released from cotemporaneous mantle-derived dykes.

4.4.5 Gold Transport and Deposition

Previous studies have shown that Au can be transported in the form of Cl^- and HS^- complexes (Robb 2007). In the Jiaodong Peninsula, gold is usually accompanied by sulfide, particularly pyrite. Wen et al. (2015) assumed that gold bisulfide complexes (e.g., $Au(HS)_2^-$ Benning and Seward 1996; Pokrovski et al. 2009) were the most likely species to transport gold. Both fluid-rock interactions and fluid immiscibility can lead to the decomposition of $Au(HS)_2^-$ complexes and depositions of gold (Naden and Shepherd 1989; Mikucki 1998; Cox et al. 1995; Qiu et al. 2017). Compared study of the Linglong and Jiaojia gold deposits in the Jiaodong Peninsula, Guo et al. (2017) assumed that fluid-rock interactions occurred in both types of gold deposition, but fluid immiscibility was more important in the Linglong-type ores. Wen et al. (2015) proposed that the water-rock interactions between ore-forming fluids and wall rocks played a highlighted role in the Jiaojia-type gold deposit, but phase separation or boiling occurred at the Linglong-type gold deposit.

In this study, the temperature and salinity of the ore-forming fluids in the Linglong gold deposit gradually decreased from the first to fourth stages. In addition, the ore-forming fluids of Linglong and Jiaojia gold deposits had a medium-high temperature, were CO_2-rich, and had a medium-low salinity. The CO_2-rich mantle-derived fluids may have caused water-rock interactions with the wall rocks of Lingong granites and Guojialing granodiorites, thereby potentially generating all possible types of alterations. In the Jiaodong Peninsula, hydrothermal fluids in the early stage of mineralisation had a medium-high temperature and a low salinity, and K-feldspar alteration and silicification developed. The hydrothermal fluids during the mineralisation period had a moderate temperature and low salinity, resulting in sericitisation. Therefore, we assumed that the water-rock interactions occurred at both types of gold deposits.

However, different scales of mineralisation occurred in different fault zones in the Jiaodong Peninsula. The orebodies in the Jiaojia-type gold deposits had lengths of hundreds to a few thousand metres and displayed disseminated gold with widths of 200–350 m and grades of <7 g/t Au (Song et al. 2015; Guo et al. 2017). However, the orebodies in the Ling-type gold deposits had lengths of tens to several thousands of metres and showed discrete veins of gold, with widths of more than 10 m and average grades of 10–20 g/t Au (Song et al. 2015; Guo et al. 2017). In addition, the temperatures of the ore-forming fluids decreased from the early to the late stage in the Jiaojia-type gold deposit (Fig. 4.12). However, the temperature of the ore-forming fluids increased weakly in the main mineralisation stage and then decreased in the late stage in the Linglong-type gold deposit (Fig. 4.12), which likely resulted from phase separation or boiling (Wen et al. 2015; Guo et al. 2020).

Given that Ce and Eu constitute abnormal-valence elements in REEs, under oxidation conditions, Ce^{3+} might oxidise to Ce^{4+} and become separated from other REEs to form Ce anomalies; these anomalies can be used to reflect the redox environment of the system (Guo et al. 2017; Li et al. 2020). The similar δCe value for pyrites from the Linglong (1.01–1.03) and Jiaojia (1.03–1.22) gold deposits and pyrites from both deposits lack of Ce anomalies, implied that the ore-forming environment was a relatively reducing environment (Fig. 4.13b); this finding was consistent with previous studies (Chen et al. 2014; Guo et al. 2017; Li et al. 2020). In addition, in high-temperature reducing hydrothermal fluids, Eu^{3+} is easily reduced to Eu^{2+} and enters the mineral phase, thereby causing Eu positive anomalies in the minerals; however, in a low-temperature reducing environment, Eu negative anomalies can easily occur (Chen et al. 2014; Li et al. 2020). The pyrites of the Linglong gold deposit had a larger range of δEu (0.97–1.97) than those of the Jiaojia gold deposit (0.58–1.02). This indicated that the Linglong gold deposit formed at a larger range of various medium-high temperatures than Jiaojia, which was consistent with the microthermometry results of fluid inclusions.

4.4.6 Deposit Genesis

Because the Jiaodong Peninsula is located at the intersection of the south-eastern part of the NCC, the northern part of the Yangtze Craton (YC), and the western part of the Pacific Plate, it experienced complex geological movements during the Mesozoic period, thus developing multi-cycle tectonic movement, metamorphism, magmatism, and mineralisation (Qiu et al. 2002; Zhou et al. 2002; Deng et al. 2003; Fan et al. 2005) (Fig. 1a). The nature of the SCLM of the eastern NCC changed significantly from the Palaeozoic to Late Mesozoic (Wu et al. 2006; Zheng et al. 2006; Zhang et al. 2008). Research on the Early Cretaceous mafic dykes and basalts in the eastern NCC showed that the eastern NCC SCLM was metasomatised through subduction of the Paleo-Pacific plate at ~177 Ma (Ma et al. 2014; Deng et al. 2017; Liang et al. 2018), subduction of the YC during the Late Triassic (Guo et al. 2014; Cai et al. 2015) or a cycling of delaminated thickened lower crust at approximately 185–165 Ma (Gao et al. 2008; Liu et al. 2009). Goldfarb and Santosh (2014) reported that potential fluid and metal reservoirs of the Jiaodong gold province included metamorphism of the subducted paleo-Pacific lithosphere and its overlying sediments, as well as devolatilisation of a hydrated mantle wedge or SCLM. Zhu et al. (2017) proposed that the melt/fluid originating from partial melting of the enriched lithospheric mantle strongly interacted with the lower crust to produce a magma with gold-rich fluid in the eastern NCC. Deng et al. (2019) showed that the high $\delta^{34}S$ values of pyrite in the Sanshandao gold deposit were derived from Neoproterozoic pyritic sedimentary rocks of the YC that subducted beneath the NCC mantle lithosphere during the Triassic. Wang et al. (2020) assumed that Mesozoic giant gold deposits in the eastern NCC were related to the volatile-rich components that accumulated during metasomatism and facilitated the release of gold during extensional craton destruction and mantle melting. In this study, we further proved that ore-forming fluids were mainly derived from high-oxygen, mantle-derived magma and ore-forming materials derived from a metasomatised SCLM; we also observed that metal was released from cotemporaneous mantle-derived dykes. This evidence established a close relationship between gold mineralisation and cotemporaneous mafic dykes in the Jiaodong gold province.

Geochronology has shown that gold mineralisation occurred at ~122 Ma in the Jiaodong Peninsula (Zhang et al. 2003a, b; Li et al. 2006; Hu et al. 2013; Deng et al. 2015b; Wang et al. 2015a, b; Ma et al. 2017; Bi and Zhao 2017; Cai et al. 2018; Feng et al. 2018; Zhang et al. 2020), as Rb-Sr isochron ages for pyrite in the Linglong gold deposit were 122–123 Ma (Yang and Zhou 2001). Meanwhile, the Early Cretaceous was the peak of lithospheric thinning of the eastern NCC, along with large-scale magmatism and mineralisation (Menzies et al. 1993; Gao et al. 2004; Liu et al. 2008; Li et al. 2012). Geochemical and geophysical data have shown that the lithospheric thickness of the eastern block of the NCC varied from ~200 km during the early Palaeozoic to less than 80 km during the Mesozoic–Cenozoic (Fan et al. 2000; Wu et al. 2006; Huang et al. 2006; Zhang et al. 2008; Chen et al. 2009; Xu et al. 2015). Although the timing and mechanisms of the destruction of the eastern NCC are still poorly understood, researchers have accepted that the Early Cretaceous was the peak of lithospheric thinning (Yang et al. 2007). Lithospheric thinning led to decompression and melting of the upwelling asthenospheric mantle, as shown via asthenospheric mantle-derived mafic dykes observed in the southwestern Jiaodong Peninsula (122–121 Ma; Ma et al. 2014); this thinning also led to partial melting of the metasomatised SCLM by underplating, generating basic magma with CO_2-containing mantle-derived fluids. Liang et al. (2019) demonstrated that the primary melts of lithospheric mantle-derived mafic dykes (110–130 Ma) had high H_2O contents in the Jiaodong Peninsula. Xia et al. (2013) also proved that Early Cretaceous Feixian basalts from the western Jiaodong Peninsula (north-eastern NCC) also had high H_2O contents. The gold that was initially within the hydrous magmas was likely partitioned into exsolved fluids and was likely enriched in the fluids by a factor of hundreds (Pokrovski et al. 2013). Thus, the mantle-derived mafic magma with high-H_2O contents was rich in CO_2-bearing mantle-source fluids and gold. In the middle–late Early Cretaceous, the Tan-Lu fault zone changed from compression to extension (Leech and Webb 2013; Deng et al. 2015b); this provided a channel for the migration of ore-forming fluids, produced regional stress differences, and promoted the upwelling of CO_2-bearing mantle-source fluids to the shallow crust. The mantle-derived fluids entered into first-level regional fracture mixing with meteoric water and occurred at a large scale of water-rock interactions, thus forming Jiaojia-type gold deposits. Meanwhile, the mantle-derived fluids entered secondary faults, and water-rock interactions and phase separation or boiling occurred, thereby forming Linglong-type gold deposits.

References

Andersen, T., H. Austrheim, E.A.J. Burke, and S. Elvevold. 1993. N_2 and CO_2 in deep crustal fluids: evidence from the caledonides of norway. *Chemical Geology* 108 (1): 113–132.

Andersen, T., E.A.J. Burke, and E.R. Neumann. 1990. Nitrogen-rich fluid in the upper mantle: fluid inclusions in spinel dunite from lanzarote, canary islands. *Contributions to Mineralogy Petrology* 120 (1): 20–28.

Benning, L.G., and T.M. Seward. 1996. Hydrosulphide complexing of gold(I) in hydrothermal solutions from 150 to 500 °C and 500 to 1500 bars. *Geochimica et Cosmochimica Acta* 60: 1849–1871.

Bi, S.J., and X.F. Zhao. 2017. $^{40}Ar/^{39}Ar$ dating of the Jiehe gold deposit in the Jiaodong Peninsula, eastern North China Craton: Implications for regional gold metallogeny. *Ore Geology Reviews* 86: 639–651.

Bi, X.W., R.Z. Hu, and J.T. Peng. 2004. REE and HFSE Geochemical Characteristics of Pyrites in Yao'an Gold Deposit: Tracing Ore Forming Fluid Signatures. *Bulleting of Mineralogy, Petrology, and Geochemistry* 23: 1–4. (in Chinese with English abs).

Bonev, I.K., T. Kerestedjian, R. Atanassova, and C.J. Andrew. 2002. Morphogenesis and composition of native gold in the chelopech volcanic-hosted au-cu epithermal deposit, srednogorie zone. *Bulgaria Mineralium Deposita* 37 (6–7): 614–629.

Cabri, L.J., N.S. Rudashevsky, V.N. Rudashevsky, and V.Y. Gorkovetz. 2008. Study of native gold from the luopensulo deposit (kostomuksha area, karelia, russia) using a combination of electric pulse disaggregation (epd) and hydroseparation (hs). *Minerals Engineering* 21 (6): 463–470.

Cai, Y.C., H.R. Fan, M. Santosh, F.F. Hu, K.F. Yang, and Z. Hu. 2015. Subduction-related metasomatism of the lithospheric mantle beneath the southeastern north china craton: evidence from mafic to intermediate dykes in the northern sulu orogen. *Tectonophysics* 659: 137–151.

Cai, Y.C., H.R. Fan, M. Santosh, F.F. Hu, K.F. Yang, and X.H. Li. 2018. Decratonic gold mineralization: Evidence from the Shangzhuang gold deposit, eastern North China Craton. *Gondwana Research* 54: 1–22.

Chen, B.H., Z.L. Wang, H.L. Li, J.K. Li, J.L. Li, and G.Q. Wang. 2014. Evolution of ore fluid of the Taishang gold deposit, Jiaodong: Constraints on REE and trace element component of auriferous pyrite. *Acta Petrologica Sinica* 30: 2518–2532. (in Chinese with English abstract).

Chen, L., C. Cheng, and Z. Wei. 2009. Seismic evidence for significant lateral variations in lithospheric thickness beneath the central and western North China Craton. *Earth Planetary Science Letters* 286: 171–183.

Chen, Y.Y. 2017. *Geochemical characteristics and genesis of the Jiaojia gold deposit in Shandong Province*. China: Chang'an University. (in Chinese with English abstract).

Clemens, J.D., G. Stevens, and F. Farina. 2011. The enigmatic sources of I-type granites The peritectic connexion. *Lithos* 126: 174–181.

Cox, S.F., S.S. Sun, M.A. Etheridge, V.J. Wall, and T.F. Potter. 1995. Structural and geochemical controls on the development of turbidite-hosted gold quartz vein deposits, Wattle Cully mine, central Victoria, Australia. *Economic Geology* 90: 1722–1746.

Deng J., B.F. Gao, Q.F. Wang, and L.Q. Yang. 2005. Formation and Evolution of Ore-Forming Fluid System. *Geological Science and Technology Information* 24(01): 49–54(in Chinese with English abstract).

Deng, J., X.F. Liu, Q.F. Wang, Y. Dilek, and Y.Y. Liang. 2017. Isotopic characterization and petrogenetic modeling of Early Cretaceous mafic diking-Lithospheric extension in the North China craton, eastern Asia. *Geological Society of America Bulletin* 129: 1379–1407.

Deng J, J.G. Wang, Y.G. Wei, Z.Q. Zhang, J.Z. Lin, and S.L. Yan. 2007. Ores and Gold-Bearing Characteristics in Xiejiagou gold deposit, Shandong province. *Earth Science- Journal of China University of Geosciences* 32(3): 373–380 (in Chinese with English abstract).

Deng, J., X.F. Liu, Q.F. Wang, and R.G. Pan. 2015a. Origin of the Jiaodong-type Xinli gold deposit, Jiaodong Peninsula, China: Constraints from fluid inclusion and C-D-O-S-Sr isotope compositions. *Ore Geology Reviews* 65: 674–686.

Deng, J., C.M. Wang, L. Bagas, E.J.M. Carranza, and Y.J. Lu. 2015b. Cretaceous-Cenozoic tectonic history of the Jiaojia Fault and gold mineralization in the Jiaodong Peninsula, China: constraints from zircon U-Pb, illite K-Ar, and apatite fission track thermochronometry. *Mineralium Deposita* 50: 987–1006.

Deng, J., Q.F. Wang, M. Santosh, X.F. Liu, Y.Y. Liang, L.Q. Yang, R. Zhao, and L. Yang. 2019. Remobilization of metasomatized mantle lithosphere: A new model for the Jiaodong gold province, eastern China. *Mineralium Deposita.* https://doi.org/10.1007/s00126-019-00925-0.

Deng, J., L.Q. Yang, and Z.S. Sun. 2003. A metallogenic model of gold deposits of the Jiaodong granite-greenstone belt. *Acta Geologica Sinica-English Edition* 77: 537–546.

Du, L.T. 1996. The relationship between crust fluids and mantle fluids. *Earth Science Frontiers* 3 (3–4): 13–21. (in Chinese with English abstract).

Fan, H.R., F.F. Hu, J.H. Yang, K. Shen, and M.G. Zhai. 2005. Fliud evolution and large-scale gold metallogeny during Mesozoic tectonic transition in the eastern Shandong province. *Acta Petrologica Sinica* 21 (5): 1317–1328. (in Chinese with English abstract).

Fan, H.R., M.G. Zhai, Y.H. Xie, and J.H. Yang. 2003. Ore-forming fluids associated with granite-hosted gold mineralization at the Sanshandao deposit, Jiaodong gold province, China. *Mineralium Deposita* 38: 739–750.

Fan, W.M., H.F. Zhang, J. Baker, K.E. Jarvis, P.R.D. Mason, and M.A. Menzies. 2000. On and off the North China Craton: Where is the Archaean keel? *Journal of Petrology* 41: 933–950.

Feng, K., H.R. Fan, D.I. Groves, K.F. Yang, F.F. Hu, X. Liu, and Y.C. Cai. 2018. Geochronological and sulfur isotopic evidence for the genesis of the post-magmatic, deeply sourced, and anomalously gold-rich Daliuhang orogenic deposit, Jiaodong, China. *Mineralium Deposita.* https://doi.org/10.1007/s00126-019-00882-8.

Fu, B., T.P. Mernagh, A.M. Fairmaid, D. Phillips, and M.A. Kendrick. 2014. CH_4-N_2 in the Maldon gold deposit, central Victoria, Australia. *Ore Geology Reviews* 58: 225–237.

Gao, S., R.L. Rudnick, W. Xu, H. Yuan, Y. Liu, R.J. Walker, I.S. Puchtel, X. Liu, H. Huang, X. Wang, and J. Yang. 2008. Recycling deep cratonic lithosphere and generation of intraplate magmatism in the North China Craton. *Earth and Planetary Science Letters* 270: 41–53.

Gao, S., R.L. Rudnick, H.L. Yuan, X.M. Liu, Y.S. Liu, W.L. Xu, W.L. Ling, A. John, X.C. Wang, and Q.H. Wang. 2004. Recycling lower continental crust in the North China craton. *Nature* 432: 892–897.

Geng, X.L., S.F. Foley, Y.S. Liu, Z.C. Wang, Z.C. Hu, and L. Zhou. 2019. Thermal-chemical conditions of the north china mesozoic lithospheric mantle and implication for the lithospheric thinning of cratons. *Earth and Planetary Science Letters* 516: 1–11.

Goldfarb, R.J., and D.I. Groves. 2015. Orogenic gold: Common or evolving fluid and metal sources through time. *Lithos* 233: 2–26.

Goldfarb, R.J., and M. Santosh. 2014. The dilemma of the Jiaodong gold deposits: Are they unique? *Geoscience Frontiers* 5: 139–153.

Gong, Q., J. Deng, C. Wang, Z. Wang, and L. Zhou. 2013. Element behaviors due to rock weathering and its implication to geochemical anomaly recognition A case study on Linglong biotite granite in Jiaodong peninsula, China. *Journal of Geochemical Exploration* 128: 14–24.

Gong, Q.J., C.W. Yun, and R.H. Zhang. 2004. Physical chemistry study on the ore-forming process of Shizhuyuan tungsten-polymetallic deposit. *Earth Science Frontiers* 11 (4): 617–625. (in Chinese with English abstract).

Graupner, T., U. Kempe, E. Dombon, O. Patzold, O. Leeder, and E.T.C. Spooner. 1999. Fluid regime and ore formation in the tungsten (-yttrium) deposits of Kyzyltau (Mongolian Altai), evidence for fluid variability in tungsten-tin ore systems. *Chemical Geology* 154: 21–58.

Griffin, W.L., S.Y. O'reilly, J.C. Afonso, and G.C. Begg. 2009. The composition and evolution of lithospheric mantle: A re-evaluation and its tectonic implications. *Journal of Petrology* 50 (7): 1185–1204.

Guo, J.T., F. Guo, C.Y. Wang, and C.W. Li. 2013a. Crustal recycling processes in generating the early Cretaceous Fangcheng basalts, North China Craton: New constraints from mineral chemistry, oxygen isotopes of olivine and whole-rock geochemistry. *Lithos* 170–171: 1–16.

Guo, P., M. Santosh, and S.R. Li. 2013b. Geodynamics of gold metallogeny in the Shandong Province, NE China an integrated geological, geophysical and geochemical perspective. *Gondwana Research* 24: 1172–1201.

Guo, F., W.M. Fan, C.W. Li, C.Y. Wang, H.X. Li, L. Zhao, and J.Y. Li. 2014. Hf-Nd-O isotopic evidence for melting of recycled sediments beneath the Sulu Orogen, North China. Chemical Geology 243–258.

Guo, L.N., J. Deng, L.Q. Yang, Z.L. Wang, S.R. Wang, Y.J. Wei, and B.H. Chen. 2020. Gold deposition and resource potential of the Linglong gold deposit, Jiaodong Peninsula: Geochemical comparison of ore fluids. *Ore Geology Reviews* 120: 103434.

Guo, L.N., R.J. Goldfarb, Z.L. Wang, R.H. Li, B.H. Chen, and J.L. Li. 2017. A comparison of Jiaojia-and Linglong-type gold deposit ore-forming fluids: do they differ? *Ore Geology Reviews* 88: 511–533.

Haas, J.R., E.L. Shock, and D.C. Sassani. 1995. Rare earth elements in hydrothermal systems: Estimates of standard partial molal thermodynamic properties of aqueous complexes of the rare earth elements at high pressures and temperatures. *Geochimica et Cosmochimica Acta* 59: 4329–4350.

Henderson, P. 1984. *Rare earth element geochemistry*, 1–510. Amsterdam: Elsevier Science Publishers.

Hronsky, J.M.A., D.I. Groves, R.R. Loucks, and G.C. Begg. 2012. A unified model for gold mineralisation in accretionary orogens and implications for regional-scale exploration targeting methods. *Mineralium Deposita* 47: 339–358.

Hu, F.F., H.R. Fan, X.H. Jiang, X.C. Li, K.F. Yang, and T. Mernagh. 2013. Fluid inclusions at different depths in the Sanshandao gold deposit, Jiaodong Peninsula, China. *Geofluids* 13: 528–541.

Huang, J., Y. Zheng, Z. Zhao, Y. Wu, J. Zhou, and X. Liu. 2006. Melting of subducted continent: element and isotopic evidence for a genetic relationship between Neoproterozoic and Mesozoic granitoids in the Sulu orogen. *Chemical Geology* 229 (4): 227–256.

Hugh, P., and J.R. Taylor. 1974. The application of oxygen and hydrogen isotope studies to problems of hydrothermal alteration and deposition. *Economic Geology* 69: 843–883.

Huraiova, M., J. Dubessy, and P. Konecny. 1991. Composition and oxidation state of upper mantle fluids from southern Slovakia. *Plinius* 5: 110–111.

Jahn, B.M., F.Y. Wu, C.H. Lo, and C.H. Tsai. 1999. Crust–mantle interaction induced by deep subduction of the continental crust: Geochemical and Sr–Nd isotopic evidence from post-collisional mafic–ultramafic intrusions of the northern Dabie complex, central China. *Chemical Geology* 157: 119–146.

Jiang, X.H., H.R. Fan, F.F. Hu, K.F. Yang, T.G. Lan, X.L. Zheng, and N.X. Jin. 2011. Comparative studies on fluid inclusion in different depth and ore genesis of the Sanshandao gold deposit, Jiaodong Peninsula. *Acta Petrologica Sinica* 27 (5): 1327–1340. (in Chinese with English abstract).

Keppler, H. 1996. Constraints from partitioning experiments on the composition of subduction-zone fluids. *Nature* 380: 237–240.

Lan, T.G., H.R. Fan, K.F. Yang, Y.C. Cai, B.J. Wen, and W. Zhang. 2015. Geochronology, mineralogy and geochemistry of alkali-feldspar granite and albite granite association from the Changyi area of Jiao-Liao-Ji Belt: Implications for Paleoproterozoic rifting of eastern North China Craton. *Precambrian Research* 266: 86–107.

Larizzatti, J.H., S.M.B. Oliveira, and C.R.M. Butt. 2008. Morphology and composition of gold in a lateritic profile, Fazenda Pison "Garimpo", Amazon, Brazil. *Journal of South American Earth Sciences* 25: 359–376.

Leech, M.L., and L.E. Webb. 2013. Is the HP–UHP Hong'an–Dabie–Sulu orogen a piercing point for offset on the Tan–Lu fault? *Journal of Asian Earth Sciences* 63: 112–129.

Li, J., M.C. Song, J.L. Liang, M.Y. Jiang, S.Y. Li, Z.J. Ding, and F. Su. 2020. Source of ore-forming fluids of the Jiaojia deeply-seated gold deposit: Evidences from trace elements and

sulfur-helium-argon isotopes of pyrite. *Acta Petrologica Sinica* 36: 297–313. (in English with abstract).
Li, S.X., C.C. Liu, and Y.H. An. 2007. *Geology of Jiaodong Gold Deposit*. Beijing: Geological Publishing House. (in Chinese with English abstract).
Li, T.G. 1985. Study on the characteristics of gold and silver occurrence in Sanshandao gold deposit. *Gold* 01: 20–24. (in Chinese with English abstract).
Li, X.C., H.R. Fan, M. Santosh, F.F. Hu, K.F. Yang, and T.G. Lan. 2013. Hydrothermal alteration associated with Mesozoic granite-hosted gold mineralization at the Sanshandao deposit, Jiaodong Gold Province, China. *Ore Geology Reviews* 403–421.
Li, J.W., S.J. Bi, D. Selby, L. Chen, P. Vasconcelos, T. David, M.F. Zhou, X.F. Zhao, Z.K. Li, and H.N. Qiu. 2012. Giant Mesozoic gold provinces related to the destruction of the North China craton. *Earth and Planetary Science Letters* 107: 437–479.
Li, J.W., P. Vasconcelos, M.F. Zhou, X.F. Zhao, and C.Q. Ma. 2006. Geochronology of the Pengjiakuang and Rushan gold deposits, eastern Jiaodong Gold Province, northeastern China: Implications for regional mineralization and geodynamic setting. *Economic Geology* 101: 1023–1038.
Li, L., M. Santosh, and S.R. Li. 2015. The'Jiaodong type' gold deposits: Characteristics, origin and prospecting. *Ore Geology Reviews* 589–611.
Li, Y.L., H.F. Zhang, J.H. Guo, and C.F. Li. 2017. Petrogenesis of the Huili Paleoproterozoic leucogranite in the Jiaobei Terrane of the North China Craton: A highly fractionated albite granite forced by K-feldspar fractionation. *Chemical Geology* 450: 165–182.
Liang, Y.Y., X.F. Liu, G.J. Li, and H.W. Li. 2014. Petrogenesis and Connection with Gold Deposits of Dikes in Jiaodong Peninsula, Eastern of North China Craton: Constrains on Geochronology and Sr-Nd-Pb Isotope. *Geological Science and Technology Information* 33 (3): 10–24. (in Chinese with English abstract).
Liang, Y.Y., J. Deng, X.F. Liu, Q.F. Wang, Y. Ma, T.X. Gao, and L.H. Zhao. 2019. Water contents of Early Cretaceous mafic dykes in the Jiaodong Peninsula, Eastern North China Craton: Insights into an enriched lithospheric mantle source metasomatized by Paleo-Pacific Plate subduction-related fluids. *Journal of Geology* 127: 262–343.
Liang, Y.Y., J. Deng, X.F. Liu, Q.F. Wang, C. Qin, Y. Li, Y. Yang, and M. Zhou. 2018. Major and trace element, and Sr isotope compositions of clinopyroxene phenocrysts in mafic dykes on Jiaodong Peninsula, southeastern North China Craton: Insights into magma mixing and source metasomatism. *Lithos* 302–303: 480–495.
Litvinenko, I.S. 2009. The conditions of existence and typomorphism of native gold in ores of the Degdekanskoe deposit (northeastern Russia) in black-shale strata. *Russian Geology and Geophysics* 50: 535–540.
Liu, J.C., X.F. Li, Y.F. Liu, B.Y. Wang, N.Z. Mi, and X. Zhang. 2010a. Mineralogical characteristics of telluride and their precipitation mechanism in the Jinqingding gold deposit, eastern Shandong, China. *Geological Bulletin of China* 29 (09): 1319–1328. (in Chinese with English abstract).
Liu, J.J., G.J. Mao, S.H. Wu, G.Z. Liu, Y.F. Liao, W.J. Zheng, S.G. Hua, and L.X. Yue. 2010b. Discovery and significance of native gold in the Zhaishang gold deposit, southern Gansu, China. *Geological Bulletin of China* 29 (1): 115–123. (in Chinese with English abstract).
Liu, R.G., and Z.R. Zhang. 1996. The mechanism of microgranular disseminated type gold during metallogenic process. *Geology-geochemistry* 6: 60–62. (in Chinese with English abstract).
Liu, S., R. Hu, S. Gao, C. Feng, B. Yu, G. Feng, Y. Qi, T. Wang, and I.M. Coulson. 2009. Petrogenesis of Late Mesozoic mafic dykes in the Jiaodong Peninsula, eastern North China Craton and implications for the foundering of lower crust. *Lithos* 113: 621–639.
Liu, J.M., J. Ye, J.H. Xu, J.G. Sun, and K. Shen. 2003. C-O and Sr–Nd isotopic geochemistry of carbonate minerals from gold deposits in East Shandong, China. *Acta Petrologica Sinica* 19: 775–784. (in Chinese with English abstract).
Liu, Y.S., S. Gao, P.B. Kelemen, and W. Xu. 2008. Recycled crust controls contrasting source compositions of Mesozoic and Cenozoic basalts in the North China Craton. *Geochimica et Cosmochimica Acta* 72: 2349–2376.

Lorand, J.P., A. Luguet, and O. Alard. 2013. Platinum-group element systematics and petrogenetic processing of the continental upper mantle: A review. *Lithos* 164: 2–21.
Lu, H.Z., H.R. Fan, P. Ni, G.X. Ou, K. Shen, and W.H. Zhang. 2004. *Fluid inclusions*, 1–487. Beijing: Science Press.
Ma, D.S. 2009. Progress in Research on Geochemistry of Tungsten. *Geological Journal of China Universities* 15 (01): 19–34. (in Chinese with English abstract).
Ma, L., S.Y. Jiang, and A. Hofman. 2014. Lithospheric and asthenospheric sources of lamprophyres in the Jiaodong Peninsula: A consequence of rapid lithospheric thinning beneath the North China Craton? *Geochimica et Cosmochimica Acta* 124: 250–271.
Ma, L., S.Y. Jiang, B.Z. Dai, Y.H. Jiang, M.L. Hou, W. Pu, and B. Xu. 2013. Multiple sources for the origin of Late Jurassic Linglong adakitic granite in the Shandong Peninsula, eastern China: zircon U-Pb geochronological, geochemical and Sr-Nd-Hf isotopic evidence. *Lithos* 162: 251–263.
Ma, W.D., H.R. Fan, X. Liu, F. Pirajno, F.F. Hu, K.F. Yang, Y.H. Yang, W.G. Xu, and P. Jiang. 2017. Geochronological framework of the Xiadian gold deposit in the Jiaodong province, China: Implications for the timing of gold mineralization. *Ore Geology Reviews* 86: 196–211.
Mao, J.W., F. Pirajno, and N. Cook. 2011. Mesozoic metallogeny in East China and corresponding geodynamic settings—An introduction to the special issue. *Ore Geology Reviews* 43: 1–7.
Mao, J.W., Y.T. Wang, H.M. Li, F. Pirajno, C.Q. Zhang, and R.T. Wang. 2008. The relationship of mantle-derived fluids to gold metallogenesisin the Jiaodong Peninsula Evidence from D-O–C–S isotope systematics. *Ore Geology Reviews* 33: 361–381.
Mao, J.W., H.M. Li, Y.T. Wang, C.Q. Zhang, and R.T. Wang. 2005. The relationship between mantle-derived fluid and gold ore-formation in the Eastern Shandong Peninsula: Evidences from D-O–C–S isotopes. *Acta Geologica Sinica* 79: 839–857. (in Chinese with English abstract).
Mcdonough, WF. 2003. Compositional model for the earth' core. In: *The mantle and core, treatise on Geochemistry*, vol. 2, ed. R.W. Carlson, 547–568. Amsterdam: Elsevier.
Menzies, M.A., W. Fan, and M. Zhang. 1993. Palaeozoic and Cenozoic lithoprobes and the loss of > 120 km of Archaean lithosphere, Sino-Korean craton, China. *Geological Society, London, Special Publications* 76: 71–81.
Mernagh, T.P. 2001. A fluid inclusion study of the Fosterville Mine: a turbidite-hosted gold field in the Western Lachlan Fold Belt, Victoria, Australia. *Chemical Geology* 173: 91–106.
Mikucki, E.J. 1998. Hydrothermal transport and depositional processes in Archean lode-gold systems: A review. *Ore Geology Reviews* 13: 307–321.
Mills, S.E., A.G. Tomkins, R.F. Weinberg, and H.R. Fan. 2015a. Implications of pyrite geochemistry for gold mineralisation and remobilisation in the Jiaodong gold district, northeast China. *Ore Geology Reviews* 150–168.
Mills, S.E., A.G. Tomkins, R.F. Weinberg, and H.R. Fan. 2015b. Anomalously silver-rich vein-hosted mineralisation in disseminated-style gold deposits, Jiaodong gold district, China. *Ore Geology Reviews* 68: 127–141.
Moh, G.H. 1980. Ore syntheses, phase equilibria studies and applications. *Neues Jahrbuch for Mineralogie. Abhandlungen* 139: 114–154.
Naden, J., and T.J. Shepherd. 1989. Role of methane and carbon dioxide in gold depositions. *Nature* 342: 793–795.
Nakagawal, M., M. Santoshl, C.G. Nambiar, and C. Matsubara. 2005. Morphology and chemistry of placer gold from Attappadi Valley, Southern India. *Gondwana Research* 8 (2): 213–222.
Ohmoto, H. 1986. Stable isotope geochemistry of ore deposits. *Reviews in Mineralogy* 16: 491–559.
Oreskes, N., and M.T. Einaudi. 1990. Origin of rare earth element-enriched hematite breccias at the Olympic Dam Cu-U-Au-Ag deposit, Roxby Downs, South Australia. *Economic Geology* 85: 1–28.
Pettke, T., R. Frei, J.D. Kramers, and I.M. Villa. 1997. Isotope systematics in vein gold from Brusson, Vald' Ayas (NW Italy) 2. (U + Th)/He and K/Ar in native Au and its fluid inclusions. *Chemical Geology* 135: 173–187.
Pokrovski, G.S., A.Y. Borisova, and A.Y. Bychkov. 2013. Speciation and transport of metals and metalloids in geological vapors. *Reviews in Mineralogy and Geochemistry* 76: 165–218.

Pokrovski, G.S., B.R. Tagirov, J. Schott, J.-L. Hazemann, and O. Proux. 2009. A new view on gold speciation in sulfur-bearing hydrothermal fluids from in situ X-ray absorption spectroscopy and quantum-chemical modeling. *Geochimica et Cosmochimica Acta* 73: 5406–5427.

Qiu, K.F., E. Marsh, H.C. Yu, K. Pfaff, C. Gulbransen, Z.Y. Gou, and N. Li. 2017. Fluid and metal sources of the Wenquan porphyry molybdenum deposit, Western Qinling, NW China. *Ore Geology Reviews* 86: 459–473.

Qiu, K.F., H.C. Yu, M.Q. Wu, J.Z. Geng, X.K. Ge, Z.Y. Gou, and R.D. Taylor. 2019. Discrete Zr and REE mineralization of the Baerzhe rare-metal deposit, China. *American Mineralogist* 104: 1487–1502.

Qiu, Y.M., D.I. Groves, N.J. Mcnaughton, L.G. Wang, and T.H. Zhou. 2002. Nature, age, and tectonic setting of granitoid-hosted, orogenic gold deposits of the Jiaodong Peninsula, eastern North China Craton, China. *Mineralium Deposita* 37: 283–305.

Robb, L. 2007. *Introduction to Ore-forming processes*. England: Black-well Publishing.

Roilison, H.R. 1993. Using geochemical data: Evaluation, Presentation, Interpretation. Longman Group UK Ltd. 275.

Rowland, J.V., and R.H. Sibson. 2004. Structural controls on hydrothermal flow in a segmented rift system, Taupo Volcanic Zone, New Zealand. *Geofluids* 4: 259–283.

Rudnick, R.L., and S. Gao. 2003. Composition of the continental crust. *Treatise on Geochemistry* 3: 659.

Shen, Y.K., T. Guo, Y.Q. Yang, Z.L. Chen, C.S. Wei, and H.S. Sun. 2016. Discovery of biotite monzolite and Ar-Ar thermochronology significance in Linglong gold field. *Jouranl of Geomechanics* 22: 778–793. (in Chinese with English abstract).

Sheppard, S.M.F. 1986. CHracterization and isotopic variations in natural waters. *Review in Mineralogy* 16: 165–183.

Song, M.C., S.Z. Li, M. Santosh, S. Zhao, S. Yu, P.H. Yi, S.X. Cui, G.X. Lv, J.X. Xu, Y.X. Song, and M.L. Zhou. 2015. Types, characteristics and metallogenesis of gold deposits in the Jiaodong Peninsula, Eastern North China Craton. *Ore Geology Reviews* 65: 612–625.

Southam, G., and T.J. Beveridge. 1996. The occurrence of sulfur and phosphorus within bacterially derived crystalline and pseudocrystalline octahedral gold formed in vitro. *Geochimica et Cosmochimica Acta* 22: 4369–4376.

Sun, S.S., and W.F. McDonough. 1989. Chemical and isotopic systematics of oceanic basalts: implications for mantle composition and processes. In: A.D. Saunders, and M.J. Norry (Eds.), *Magmatism in the Oceanic Basins: Geological Society Special Publications* (pp. 313–345). London.

Tang, H.Y., J.P. Zheng, and C.M. Yu. 2008. Age and composition of the Rushan intrusive complex in the northern Sulu orogen, eastern China: petrogenesis and lithospheric mantle evolution. *Geological Magazine* 1: 1–17.

Tan, J., J. Wei, A. Audétat, and T. Pettke. 2012. Source of metals in the Guocheng gold deposit, Jiaodong Peninsula, North China Craton: Link to early Cretaceous mafic magmatism originating from Paleoproterozoic metasomatized lithospheric mantle. *Ore Geology Reviews* 48: 70–87.

Tan, J., J.H. Wei, Y.J. Li, L.B. Fu, H.M. Li, W.J. Shi, and N. Tian. 2015. Origin and geodynamic significance of fault-hosted massive sulfide gold deposits from the Guocheng-Liaoshang metallogenic belt, eastern Jiaodong Peninsula: Rb-Sr dating, and H-O-S-Pb isotopic constraints. *Ore Geology Reviews* 65: 687–700.

Tassara, S., J.M. González-Jiménez, M. Reich, M.E. Schilling, D. Morata, G. Begg, E. Saunders, W.L. Griffn, S.Y. O'Reilly, M. Grégoire, F. Barra, and A. Corgne. 2017. Plume-subduction interaction forms large auriferous provinces. *Nature Communications* 8: 843.

Wang, C.M., J. Deng, M. Santosh, J.M.C. Emmanuel, Q.J. Gong, C.Y. Guo, R. Xia, and X.R. Lai. 2015a. Timing, tectonic implications and genesis of gold mineralization in the Xincheng gold deposit, China: C-H-O isotopes, pyrite Rb-Sr and zircon fission track thermochronometry. *Ore Geology Reviews* 65: 659–673.

Wang F.N. 1987. Tungsten deposits related to gegional metamorphism and ore-controlling factors. *Geological Review* 33(5): 417–425 (in Chinese with English abstract).

Wang, L.G., Q. Qiu, N.J. McNaughton, D.I. Groves, Z. Luo, J. Huang, L. Miao, and Y. Liu. 1998. Constraints on crustal evolution and gold metallogeny in the northwestwern Jiaodong Peninsula, China, from SHRIMP U–Pb zircon studies of granitoids. *Ore Geological Review* 13: 275–291.

Wang, Z.L., L.Q. Yang, L.N. Guo, E. Marsh, J.P. Wang, Y. Liu, C. Zhang, R.H. Li, L. Zhang, X.L. Zheng, and R.X. Zhao. 2015b. Fluid immiscibility and gold deposition in the Xincheng deposit, Jiaodong Peninsula, China: Afluid inclusion study. *Ore Geology Reviews* 65: 701–717.

Wang, Q.F., J. Deng, G.J. Li, J.Y. Liu, C. Li, and E.M. Ripley. 2019a. Geochronological, petrological, and geochemical studies of the Daxueshan magmatic Ni-Cu sulfide deposit in the Tethyan Orogenic Belt, Southwest China. *Economic Geology* 113: 1307–1332.

Wang, Z.C., H. Cheng, K.Q. Zong, X.L. Geng, Y.S. Liu, J.H. Yang, F.Y. Wu, H. Becker, S. Foley, and C.Y. Wang. 2019b. Metasomatized lithospheric mantle for Mesozoic giant gold deposits in the North China Craton. *Geology*. https://doi.org/10.1130/G46662.1/4880265/g46662.

Wang, Q.F., D.I. Groves, J. Deng, H.J. Li, L. Yang, and C.Y. Dong. 2020a. *Evolution of the Miocene Ailaoshan orogenic gold deposits, southeastern Tibet, during a complex tectonic history of lithosphere-crust interaction*. Miner: Deposita. https://doi.org/10.1007/s00126-019-00922-3.

Wen, B.J., H.R. Fan, M. Santosh, F.F. Hu, F. Pirajno, and K.F. Yang. 2015. Genesis of two different types of gold mineralization in the Linglong gold field, China: Constrains from geology, fluid inclusions and stable isotope. *Ore Geology Reviews* 65: 643–658.

Wood, S.A., and D. Vlassopoulos. 1989. Experimental determination of the hydrothermal solubility and speciation of tungsten at 500 °C and 1 kbar. *Geochimica et Cosmochimica Acta* 53: 303–312.

Wu, F.Y., R.J. Walker, Y.H. Yang, H.L. Yuan, and J.H. Yang. 2006. The chemical-temporal evolution of lithospheric mantle underlying the North China Craton. *Geochimica et Cosmochimica Acta* 70: 5013–5034.

Wu, M.L., G.C. Zhao, M. Sun, S.Z. Li, P. Yuktam, P.R. Eizenhöefer, Y.H. He. 2014. Zircon U–Pb geochronology and Hf isotopes of major lithologies from the Jiaodong Terrane: Implications for the crustal evolution of the Eastern. *Lithos* 71–84.

Xia L.X., W.F. Liu, and N.A. Tang. 1990. Abundances of fluorine, chlorine, sulfur and phosphorus and their evolution in volcanic magmas from Wudalianchi. *Geochimica* (03): 197–206 (in Chinese with English abstract).

Xia, Q.K., J. Liu, S.C. Liu, I. Kovács, M. Feng, and L. Dang. 2013. High water content in mesozoic primitive basalts of the North China Craton and implications on the destruction of cratonic mantle lithosphere. *Earth Planetary Science Letters* 361: 85–97.

Xu, Y., H. Zeyen, T.Y. Hao, M. Santosh, Z.W. Li, S. Huang, and J. Xing. 2015. Lithospheric structure of the North China Craton: Integrated gravity, geoid and topography data. *Gondwana Research* 34: 315–323.

Xu Y.S., B.R. Zhang, and Y.W. Han. 1992. An experimental study on the partitioning of tungsten between aqueous fluids and silicate melts. *Geochimica* (03): 273–281 (in Chinese with English abstract).

Yang, C.X., and M. Santosh. 2019. Ancient deep roots for Mesozoic world-class gold deposits in the North China Craton: an integrated genetic perspective. *Geoscience Frontiers*. https://doi.org/10.1016/j.gsf.2019.03.002.

Yang H.Y., S.H. Wang, X.L. Song, H.D. Pan, and P.C. Ma. 2011. Gold occurrence of Jiaojia gold mine in Shandong province. *Transactions of Nonferrous Metals Society of China* 21(9): 2072–2077.

Yang, J.H., F.Y. Wu, S.L. Chung, C.H. Lo, S.A. Wilde, and G.A. Davis. 2007. Rapid exhumation and cooling of the Liaonan metamorphic core complex: Inferences from $^{40}Ar/^{39}Ar$ thermochronology and implications for Late Mesozoic extension in the eastern North China Craton. *Geological Society of America Bulletin* 119: 1405–1414.

Yang, J.H., and X.H. Zhou. 2001. Rb-Sr, Sm-Nd, and Pb isotope systematics of pyrite: Implications for the age and genesis of lode gold deposits. *Geology* 8: 711–714.

Yang, K.F., H.R. Fan, M. Santosh, F.F. Hu, S.A. Wilde, and T.G. Lan. 2012. Reactivation of the Archean lower crust: Implications for zircon geochronology, elemental and Sr-Nd-Hf isotopic geochemistry of Late Mesozoic granitoids from northwestern Jiaodong terrane, the North China Craton. *Lithos* 146–147: 112–127.

Yang, L.Q., J. Deng, L.N. Guo, Z.L. Wang, X.Z. Li, and J.L. Li. 2016a. Origin and evolution of ore fluid, and gold deposition processes at the giant Taishang gold deposit, Jiaodong Peninsula, eastern China. *Ore Geology Reviews* 72: 582–602.
Yang, L.Q., J. Deng, Z.L. Wang, L.N. Guo, R.H. Li, D.I. Groves, L.V. Danyushevsky, C. Zhang, X.L. Zheng, and H. Zhao. 2016b. Relationships between gold and pyrite at the Xincheng gold deposit, Jiaodong Peninsula, China: Implications for gold source and deposition in a brittle epizonal environment. *Economic Geology* 111: 105–126.
Yang Q.Y., J.F. Shen, S.R. Li, M. Santosh, Z.H. Luo, and Y. Liu. 2013. Oxygen, boron, chromium and niobium enrichment in native Au and Ag grains: A case study from the Linglong gold deposit, Jiaodong, eastern China. *Journal of Asian Earth Sciences* 62(SI): 537–546.
Yin, Q.Z., S.B. Jacobsen, K. Yamashita, J. Blichert-Toft, P. Telouk, and F. Albarede. 2002. A short timescale for terrestrial planet formation from Hf-W chronometry of meteorites. *Nature* 418: 949–951.
Yu Y.M., C.H. Gammons, and A.E. Williams-Jones. 1997. A possible mechanism of Au settlement: disproportionated reaction. *Science in China (Series D)* 27(5): 419– 424 (in Chinese with English abstract).
Zeng J.N., Y.X. Fan. 2002. Experiments of Mechanisms of Gold Precipitated by Fluid-Mixing Function. *Earth Science- Journal of China University of Geosciences*27(1): 41–45 (in Chinese with English abstract).
Zhang H.F., S.R. Li, M.G. Zhai, and J.H. Guo. 2006. Crust uplift and its implications in the Jiaodong Peninsula, eastern China. *Acta Petrologica Sinca* 22(02): 285–295 (in Chinese with English abstract).
Zhang H., M. Sun, X. Zhou, W. Fan, M. Zhai, and J. Yin. 2002. Mesozoic lithosphere destruction beneath the North China Craton: evidence from major-, trace-element and Sr-Nd-Pb isotope studies of Fangcheng basalts. *Contributions to Mineralogy and Petrology* 144(2): 241–254.
Zhang H.F., and S. Gao. 2012. *Geochemistry* (pp. 1-410). Beijing: Geology press (in Chinese with English abstract).
Zhang, H.F., S.L. Goldstein, X.H. Zhou, M. Sun, J.P. Zheng, and Y. Cai. 2008. Evolution of subcontinental lithospheric mantle beneath eastern China: Re-Os isotopic evidence from mantle xenoliths in Paleozoic kimberlites and Mesozoic basalts. *Contributions to Mineralogy and Petrology* 155: 271–293.
Zhang, J., Z.F. Zhao, F.Y. Zheng, X.M. Liu, and L.W. Xie. 2012. Zircon Hf–O isotope and whole-rock geochemical constraints on origin of postcollisional mafic to felsic dykes in the Sulu orogen. *Lithos* : 136–139, 225–245.
Zhang, L., R.F. Weinberg, L.Q. Yang, D.I. Groves, S.X. Sai, E. Matchan, D. Phillips, B.P. Kohn, D.P. Miggins, Y. Liu, and J. Deng. 2020. Mesozoic orogenic gold mineralization in the Jiaodong Peninsula, China: A focused event at 120 ± 2 Ma during cooling of pregold granite intrusions. *Economic Geology* 115: 415–441.
Zhang, L.C., S.Y. Chao, Q.D. Zeng, and G.M. Li. 2003a. $^{40}Ar/^{39}Ar$ and Rb-Sr isochron dating of the gold deposits on northern margin of the Jiaolai Basin, Shandong, China. *Science in China (Series D)* 46: 708–718.
Zhang, X.O., P.A. Cawood, S.A. Wilde, R.Q. Liu, H.L. Song, W. Li, and L.W. Snee. 2003b. Geology and timing of mineralization at the Cangshang gold deposit, northwestern Jiaodong Peninsula, China. *Mineralium Deposita* 38: 141–153.
Zhao, H.S., Q.F. Wang, D.I. Groves, and J. Deng. 2019. A rare Phanerozoic amphibolite-hosted gold deposit at Danba, Yangtze Craton, China: significance to fluid and metal sources for orogenic gold systems. *Mineralium Deposita* 54 (1): 133–152.
Zheng, J.P., W.L. Griffin, S.Y. O'Reilly, J.L.T. Yang, M. Zhang, R.Y. Zhang, and J.G. Liou. 2006. Mineral chemistry of peridotites from Paleozoic, Mesozoic and Cenozoic lithosphere: Constraints on mantle evolution beneath Eastern China. *Journal of Petrology* 47: 2233–2256.
Zhou, T.H., R.J. Goldfarb, and G.N. Phillips. 2002. Tectonics and distribution of gold deposits in China-an overview. *Mineralium Deposita* 37: 249–282.

Zhu, R.X., H.R. Fan, J.W. Li, Q.R. Meng, S.R. Li, and Q.D. Zeng. 2015. Decratonic gold deposits. *Science China Earth Sciences* 58: 1523–1537.

Zhu, R.X., H.F. Zhang, G. Zhu, Q.R. Meng, H.R. Fan, J.H. Yang, F.Y. Wu, Z.Y. Zhang, and T.Y. Zheng. 2017. Craton destruction and related resources. *International Journal of Earth Sciences* 106: 2233–2257.

The manufacturer's authorised representative in the EU is Springer Nature Customer Service Centre GmbH, Europaplatz 3, 69115 Heidelberg, Germany. If you have any concerns regarding our products, please contact ProductSafety@springernature.com

Printed and bound by CPI Group (UK) Ltd, Croydon, CR0 4YY
30/09/2026
02225289-0002